Electrical Machines

Electrical Machines
An Introduction to Principles and Characteristics

J. D. EDWARDS

Lecturer in Electrical Engineering
The University of Sussex

INTERNATIONAL TEXTBOOK COMPANY

Published by
International Textbook Company Limited
A member of the Blackie Group
450 Edgware Road
London W2 1EG

First published 1973
Reprinted 1976

ISBN 0 7002 0267 6

Text set in 10/12 pt. IBM Press Roman, printed by photolithography,
and bound in Great Britain at The Pitman Press, Bath

Contents

Preface

The aim of this book is to present the principles of electrical machines in a
form accessible to the non-specialist, yet with sufficient rigour to provide a basis
for more advanced studies. This has been achieved by abandoning the classical
approach, in which each type of machine is studied in isolation with considerable
emphasis on the detailed structure of machine windings.

Electrical machines form only part of a large class of electromechanical
devices which all utilize the same basic physical principles, and recognition of
this fact has led to unified methods of analysis based on energy and circuit
concepts. These methods provide a radical alternative to the classical study of
machines, but they have two disadvantages for an introductory treatment: they
involve abstract concepts and extensive analysis; and they progress from the general
to the particular when studying individual machine types. The author has therefore chosen a direct physical approach in which the action of each type of machine
is studied in terms of the interaction of currents and magnetic fields. Unity is
provided by a common method of analysis, and an important feature of the
classical treatment is retained: the formation of a clear physical picture of the
operating principles of each type of machine.

The treatment is fully quantitative, and the requirements of rigour and
simplicity have been met by taking the simplest model of a machine which will
demonstrate the essential features of its operation. Departures from the ideal are
mentioned only briefly, since these form the subject of specialized study. Likewise the varieties of practical windings are not treated; but the functions of
machine windings are fully explained. Particular attention has been given to the
treatment of a.c. machines, since these give students the most difficulty. The
sinusoidally distributed magnetic field is taken as the key concept, from which the
rotating field principle, the torque equation and the equivalent circuit may all be
derived.

In a concluding chapter there is an introduction to Kron's generalized
mathematical theory, which demonstrates the essential unity of all rotating

vii

machines. The earlier chapters provide a background of physical principles essential to a proper understanding of machines, and the power and elegance of the generalized theory can then be appreciated. Quite apart from any intellectual satisfaction it may provide, the theory is indispensable for advanced work such as the transient behaviour of a.c. machines. Since rotating electrical machines are components of most engineering systems, it seems desirable to give at least an introduction to this powerful method of analysis; but the chapter may well be omitted at a first reading.

The reader is assumed to have an elementary knowledge of electromagnetism, circuit theory, vectors, matrices and differential equations. Suitable background material is listed in the bibliography at the end of the book, along with suggestions for further reading.

I wish to thank Mr. A. Draper for permission to reproduce Figs. 2.10, 4.25 and 6.6; and the Patent Office for permission to reproduce Fig. 1.34. The originality of a textbook must lie in the presentation rather than the ideas themselves, and my debt to other writers will be evident. In particular, my treatment of the rotating magnetic field owes much to C. R. Chapman's lucid exposition.* I also wish to record my thanks to Dr. B. V. Jayawant for reading the manuscript and making valuable suggestions, to Dr. G. Williams for assistance in correcting the proofs, and to my wife for typing the manuscript.

* CHAPMAN, C. R. (1965). *Electromechanical energy conversion.* Blaisdell Publishing Co., New York.

Notation, Units and Symbols

SI units are used throughout the book, and the recommendations of the British Standards Institution [1] for unit names and symbols are followed with one exception: the ampere turn is retained in place of the ampere for the unit of magnetomotive force, on the grounds that it is useful to distinguish between current and magnetomotive force.

In general the recommendations of the Institution of Electrical Engineers [2] and the British Standards Institution [3] have been followed for symbols, abbreviations and subscripts. Symbols for physical quantities are printed in sloping type, and unit symbols in upright type. With electrical quantities, the usual convention is followed in denoting time-varying values by lower-case symbols and steady values or magnitudes by upper-case symbols.

Three kinds of vector quantity occur in the book: space phasors, time phasors and three-dimensional vectors. Since there is no possibility of confusion between time phasors and three-dimensional vectors, bold-face type is used for these quantities in accordance with normal practice, e.g. I, B. It is necessary, however, to distinguish between vectors and space phasors; space phasors are denoted by an arrow over the symbol, e.g. $\vec{B}$. Bold-face type is also used to denote matrices.

There is one area of notation in which no uniformity exists: the choice of symbols for the infinitesimal elements in line, surface and volume integrals. After careful consideration the notation adopted is that of Stratton [4], in which ds, da and dv denote the elements of path length, area and volume respectively. This is a consistent and convenient notation; it justifies the use of the subscript s for the tangential component of a vector; it releases capital letters for designating finite regions or quantities; and there is no confusion in advanced work between the symbol for area and the Poynting vector S or the vector potential A.

The following lists refer only to the usage in this book. Further information will be found in the references already cited and in Massey [5].

List of principal symbols

Symbol	Quantity	Unit	Unit Symbol
A	area	square metre	m^2
a	radius	metre	m
B, B	magnetic flux density	tesla	T
$\vec{B}$	flux density phasor	tesla	T
C	capacitance	farad	F
da, da	element of area	square metre	m^2
ds, ds	element of path length	metre	m
dv	element of volume	cubic metre	m^3
E, E	electric field strength	volt/metre	V/m
$E; E$	excitation voltage phasor; r.m.s. magnitude	volt	V
$E; E$	induced e.m.f. phasor; r.m.s. magnitude	volt	V
E, e	electromotive force	volt	V
F, F	mechanical force	newton	N
F	magnetomotive force	ampere turn	At
F_x	x-component of force	newton	N
f, f	force per unit volume	newton/metre^3	N/m^3
f	frequency	hertz	Hz
g	air-gap length	metre	m
H, H	magnetizing force	ampere turn/metre	At/m
$I; I$	current phasor; r.m.s. magnitude	ampere	A
I_0	no-load current	ampere	A
I_{0l}	loss component of I_0	ampere	A
I_{0m}	magnetizing component of I_0	ampere	A
I, i	current	ampere	A
J, J	current density	ampere/metre^2	A/m^2
J	moment of inertia	kilogram metre^2	kg m^2
j	$\pi/2$ operator, $\sqrt{(-1)}$	—	—
K	torque constant	newton metre/tesla^2	N m/T^2
K	d.c. machine constant	volt second/ampere radian / newton metre/ampere^2	V s/A rad / N m/A^2
K_a	armature constant	volt second/weber radian / newton metre/weber ampere	V s/Wb rad / N m/Wb A
K_f	field constant	weber/ampere	Wb/A

Symbol	Quantity	Unit	Unit Symbol
k	conductor density	radian^{-1}	rad^{-1}
L	self inductance	henry	H
L_m	magnetizing inductance	henry	H
l	leakage inductance	henry	H
l	length	metre	m
M	mutual inductance	henry	H
m	number of phases	—	—
N, n	number of turns	—	—
n	Steinmetz index	—	—
n	turns ratio	—	—
P	power	watt	W
P_e	electrical power	watt	W
P_m	mechanical power	watt	W
p	d/dt operator	second^{-1}	s^{-1}
p	number of pole pairs	—	—
p_e	eddy current power loss per unit volume	watt/metre^3	W/m^3
p_h	hysteresis power loss per unit volume	watt/metre^3	W/m^3
q	electric charge	coulomb	C
R	resistance	ohm	Ω
R_e	equivalent total resistance	ohm	Ω
R_l	core loss resistance	ohm	Ω
r	radius	metre	m
S	reluctance	ampere turn/weber	At/Wb
s	fractional slip	—	—
T	torque	newton metre	N m
T_m	mechanical output torque	newton metre	N m
T_θ	torque associated with angle θ	newton metre	N m
t, t	stress	newton/metre^2	N/m^2
t	time	second	s
U	magnetic potential difference	ampere turn	At
u, u	linear velocity	metre/second	m/s
$V; V$	voltage phasor; r.m.s. magnitude	volt	V
V, v	terminal voltage, electric potential difference	volt	V
W	energy, work done	joule	J
W_e	electrical energy	joule	J
W_m	magnetic stored energy	joule	J
w_h	hysteresis energy loss per unit volume	joule/metre^3	J/m^3

Symbol	Quantity	Unit	Unit Symbol
X	reactance	ohm	Ω
X_e	equivalent total reactance	ohm	Ω
X_m	mutual or magnetizing reactance	ohm	Ω
X_s	synchronous reactance	ohm	Ω
x	leakage reactance	ohm	Ω
Z, Z	impedance	ohm	Ω
α, β, γ	general angles	radian	rad
δ	load angle, angle between magnetic field axes	radian	rad
δ	depth of penetration	metre	m
ϵ	voltage regulation	–	–
η	efficiency	–	–
θ	angular displacement, rotor angle	radian	rad
Λ	permeance	weber/ampere turn	Wb/At
λ_h	Steinmetz coefficient	–	–
μ	absolute permeability $= \mu_0\mu_r$	henry/metre	H/m
μ_0	magnetic constant $= 4\pi \times 10^{-7}$	henry/metre	H/m
μ_r	relative permeability	–	–
ρ	charge per unit volume	coulomb/metre^3	C/m^3
ρ	resistivity $= 1/\sigma$	ohm metre	Ω m
σ	conductivity $= 1/\rho$	siemens/metre	S/m
τ	time constant	second	s
τ_{em}	electromechanical time constant	second	s
Φ	magnetic flux, flux per pole	weber	Wb
Φ_l	leakage flux	weber	Wb
ϕ	phase angle	radian	rad
χ	angular displacement in rotor	radian	rad
ψ	angular displacement	radian	rad
Ω, ω	angular velocity	radian/second	rad/s
ω	angular frequency $= 2\pi f$	radian/second	rad/s
ω_r	rotor angular velocity	radian/second	rad/s
ω_s	synchronous angular velocity	radian/second	rad/s

General subscripts

a	armature
av	average
f	field
m, max	maximum value
n	normal component

r	radial component
s	tangential component
1, 2	primary, secondary; stator, rotor
α, β	two-phase quantities
a, b, c	three-phase quantities
d, q	two-axis armature quantities
f, g	two-axis field quantities

Decimal prefixes

10^6	mega	M
10^3	kilo	k
10^{-2}	centi	c
10^{-3}	milli	m
10^{-6}	micro	μ

Abbreviations

a.c.	alternating current
d.c.	direct current
e.m.f.	electromotive force
m.m.f.	magnetomotive force
rev/min	revolutions per minute
rev/s	revolutions per second
r.m.s.	root-mean-square

References

[1] BRITISH STANDARDS INSTITUTION: BS 3763 (1964). *The international system (SI) units.*
[2] INSTITUTION OF ELECTRICAL ENGINEERS (1968). *Symbols and abbreviations for use in electrical and electronic engineering course.* London.
[3] BRITISH STANDARDS INSTITUTION: BS 1991 (1967). *Letter symbols, signs and abbreviations.*
[4] STRATTON, J. A. (1941). *Electromagnetic theory.* McGraw-Hill, New York.
[5] MASSEY, B. S. (1971). *Units, dimensional analysis and physical similarity.* Van Nostrand Reinhold, London.

CHAPTER 1

General Principles

1.1 Introduction

In 1820 Oersted discovered the magnetic effect of an electric current, and the first primitive electric motor was built in the following year. Faraday's discovery of electromagnetic induction in 1831 completed the foundations of electromagnetism, and the principles were vigorously exploited in the rapidly growing field of electrical engineering (Dunsheath [1]). By 1890 the main types of rotating electrical machine had been invented, and the next thirty years saw the development of many ingenious variations, along with refinement of the basic types. This was the golden age of machine development, but the field is by no means static; discoveries of a fundamental nature continue to be made, and important developments have followed the growth of other technologies such as solid-state electronics and cryogenics.

There is an enormous variety of electrical machines, but most of them are derived from the three main types treated in this book: simple d.c. machines, a.c. synchronous machines and a.c. induction machines. In spite of their superficial differences they all exploit the effects discovered by Oersted and Faraday. These two effects and their interrelation are seen most clearly when a current-carrying conductor is free to move in a magnetic field of constant intensity; this case is considered in section 1.2. But in most practical machines the conductors are not free to move; they are embedded in slots in the iron core of the machine, and forces act on the iron as well as on the conductors. It is necessary therefore to consider force production and e.m.f. generation in more general terms, which is done in sections 1.3 and 1.4. Magnetic materials form an essential part of electrical machines; some of their properties are discussed in section 1.5, and the important concept of the magnetic circuit is introduced in section 1.6.

1.2 Conductor moving in a magnetic field

When a conductor moves in a magnetic field, an e.m.f. is generated; when it carries a current in a magnetic field, a force is produced. Both of these effects may be deduced from one of the most fundamental principles of electromagnetism, and they provide the basis for a number of devices in which conductors move freely in a magnetic field. It has already been mentioned that most electrical machines employ a different form of construction, and the concepts developed in the next two sections are necessary for a proper understanding of their operation. Nevertheless the equations developed in this section for the force and the induced e.m.f. remain valid for many practical machines; this important and useful result will be justified in chapter 2.

INDUCED E.M.F. IN A MOVING CONDUCTOR

Consider a conductor moving with a velocity denoted by the vector u in a magnetic field B (Fig. 1.1). If the conductor slides along wires connected to a voltmeter, there will be a reading on the meter, showing that an e.m.f. is being

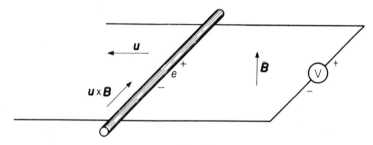

Fig. 1.1

generated in the circuit. The effect may be explained in terms of the Lorentz equation for the force on a moving charge q:

$$F = q(E + u \times B) \text{ newtons,} \qquad (1.1)$$

where q is the charge in coulombs, E the electric field strength in volts/metre, u the velocity in metres/second, and B the magnetic flux density in teslas. If the conductor is initially at rest, there will be no electric field E and no reading on the voltmeter. When the conductor is moving with velocity u, a force $qu \times B$ will act on any charged particle within the conductor, and the free charges (conduction electrons) will begin to move under the action of the force. There will be an accumulation of negative charge at one end of the conductor, leaving a surplus of positive charge at the other end; this will set up an electrostatic field E, and charge separation will continue until the force qE is exactly equal and opposite to $q\boldsymbol{u} \times \boldsymbol{B}$. The net force is then zero; there is no further motion of charge, and we have

$$\boldsymbol{E} = -\boldsymbol{u} \times \boldsymbol{B}. \tag{1.2}$$

The quantity $\boldsymbol{u} \times \boldsymbol{B}$ may be regarded as an induced electric field produced by the motion of the conductor, and this is opposed by an equal and opposite electrostatic field $\boldsymbol{E}$ produced by a distribution of electric charge. In virtue of the electrostatic field $\boldsymbol{E}$ there will be an electrostatic potential difference between the ends of the conductor given by the line integral of $\boldsymbol{E}$ along any path joining the ends PQ:

$$v = -\int_P^Q \boldsymbol{E} \cdot d\boldsymbol{s} \text{ volts}, \tag{1.3}$$

and this will be measured by a voltmeter connected between the wires. From equation 1.2 this may be written as $v = e$, where

$$e = \int_P^Q \boldsymbol{u} \times \boldsymbol{B} \cdot d\boldsymbol{s} \text{ volts}, \tag{1.4}$$

and e may be regarded as the e.m.f. induced in the conductor by its motion in the magnetic field. The integral may conveniently be taken along the axis of the conductor, which is a line of length l denoted by the vector $\boldsymbol{l} = \overrightarrow{PQ}$ If $\boldsymbol{u}, \boldsymbol{B}$ and the conductor length l are mutually perpendicular, the induced electric field $\boldsymbol{u} \times \boldsymbol{B}$ will be parallel to $\boldsymbol{l}$, and its direction is determined by the right-hand screw rule of the vector product. (If one looks in the direction of $\boldsymbol{u} \times \boldsymbol{B}, \boldsymbol{B}$ is displaced in a clockwise direction from $\boldsymbol{u}$.) If the magnetic field is uniform along the length of the conductor, we then have

$$e = \int_0^l uB\,ds = Blu \text{ volts}, \tag{1.5}$$

and the sign of e is determined from the direction of $\boldsymbol{u} \times \boldsymbol{B}$ as shown in Fig. 1.1. Equation 1.5 is commonly known as the 'flux cutting rule', or more accurately as the motional induction formula; its relation to Faraday's law of electromagnetic induction is discussed in section 1.3

Conductor resistance

Suppose that a resistance is connected in place of the voltmeter, so that a current i flows. If this current is distributed uniformly over the cross-section of the conductor, and the cross-sectional area is A, the magnitude of the current density is·

$$J = \frac{i}{A} \text{ amperes/metre}^2 \tag{1.6}$$

and its direction is along the conductor, as shown in Fig. 1.2. Since the total
force acting on unit charge is $E + u \times B$, Ohm's law for the moving conductor is

$$J = \sigma(E + u \times B), \tag{1.7}$$

where σ is the conductivity of the material. Thus the electrostatic field E must
be slightly less than the induced electric field $u \times B$ when a current is flowing, the

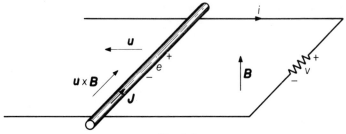

Fig. 1.2

resultant force per unit charge being just sufficient to maintain the flow of
current. The potential difference between the ends of the conductor is now
given by

$$v = -\int E \cdot ds = \int u \times B \cdot ds - \int \frac{1}{\sigma} J \cdot ds$$

$$= \int_0^l uB \, ds - \int_0^l \frac{i}{\sigma A} \, ds = Blu - \frac{li}{\sigma A} \tag{1.8}$$

$$\text{or } v = e - Ri \tag{1.9}$$

where $e = Blu$ is the induced e.m.f. in volts and $R = l/\sigma A$ is the resistance of the
conductor in ohms. The system may be represented by an equivalent circuit, as
shown in Fig. 1.3.

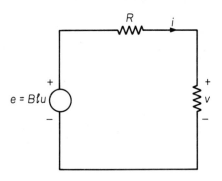

Fig. 1.3

ELECTROMAGNETIC FORCE ON A CONDUCTOR

Take the same configuration of a conductor in a magnetic field, and suppose initially that the conductor is stationary (Fig. 1.4). If a current i is flowing, there will be a flow of free charge along the conductor; let ρ be the charge per unit volume, and U the average drift velocity of the charge. This moving charge

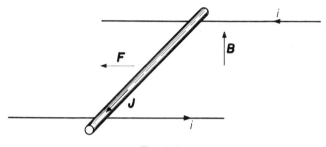

Fig. 1.4

will experience a force in a magnetic field, and from equation 1.1 the force per unit volume is

$$f = \rho U \times B \text{ newtons/metre}^3. \tag{1.10}$$

Since the free charge cannot escape from the sides, this force is transmitted to the conductor. We may express the force in terms of the current by noting that the current density J is given by

$$J = \rho U. \tag{1.11}$$

The force per unit volume is therefore

$$f = J \times B, \tag{1.12}$$

and the total force on the conductor is given by the volume integral

$$F = \int J \times B \, dv \text{ newtons}. \tag{1.13}$$

In the simple case of a uniform current density given by $J = i/A$ (equation 1.6), and a uniform magnetic flux density B perpendicular to the conductor,

$$F = \int JB \, dv = \int_0^l \frac{iB}{A} A \, ds = Bli \text{ newtons}. \tag{1.14}$$

From equation 1.12 the direction of the force is given by the vector product $J \times B$, and it is perpendicular to both the conductor and the magnetic field.

ELECTROMECHANICAL ENERGY CONVERSION

Fig. 1.5 shows the conductor connected to a voltage source v. There is a current i flowing, and the conductor is moving with a velocity u. The directions of the force F and the induced e.m.f. e are shown in the figure. Since the force F

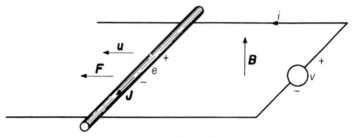

Fig. 1.5

and the velocity u are in the same direction, the conductor does mechanical work at the rate

$$P_m = F . u = Fu \text{ watts.} \tag{1.15}$$

The voltage source v is driving a current i into the circuit, and it therefore does work at the rate

$$P_e = vi \text{ watts.} \tag{1.16}$$

Since the direction of current flow is the reverse of that in Fig. 1.2, equation 1.9 becomes

$$v = e + Ri$$

$$= Blu + Ri, \tag{1.17}$$

and we also have the force equation

$$F = Bli. \tag{1.14]*}$$

Multiplying equation 1.17 by i and equation 1.14 by u gives

$$P_e = vi = Blui + Ri^2,$$

$$P_m = Fu = Bliu.$$

It follows that

$$P_e = P_m + Ri^2, \tag{1.18}$$

* Square brackets indicate that the equation, with this number, was introduced in an earlier part of the book.

showing that the electrical input power P_e is equal to the mechanical output power P_m plus the ohmic losses in the conductors; the device is acting as a motor, converting electrical energy into mechanical energy. For the current to flow in the direction shown, the applied voltage v must exceed the induced e.m.f. e; if v is smaller than e the direction of current flow is reversed, and the direction of the force F is reversed in consequence. The conductor then absorbs mechanical energy at the rate $P_m = Fu$; the voltage source likewise absorbs electrical energy at the rate $P_e = vi$, and

$$P_m = P_e + Ri^2 .$$ (1.19)

The device is then acting as a generator, converting mechanical energy into electrical energy plus ohmic losses. The process of energy conversion is reversible, and there is no fundamental difference between generator and motor action.

APPLICATIONS

Among the best known applications of the conductor in a magnetic field are the moving-coil loudspeaker (Fig. 1.6) and the moving-coil galvanometer (Fig. 1.7). In each of these devices the conductor, in the form of a coil, moves in the uniform radial field of a permanent magnet; motion of the coil is opposed by a spring, giving a displacement proportional to the coil current. The direct

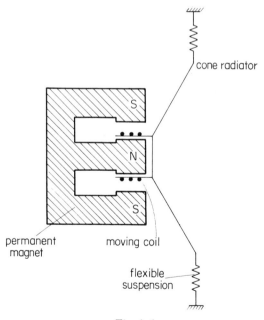

cone radiator

permanent magnet

moving coil

flexible suspension

Fig. 1.6

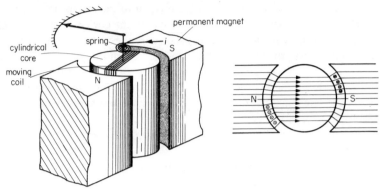

Fig. 1.7

proportionality of the force or displacement to the current makes the moving coil principle particularly useful for instrumentation. In the force-balance accelerometer, for instance, a feedback system senses the coil displacement and adjusts the current until the electromagnetic force exactly balances the acceleration reaction force. The coil current then gives a measure of the acceleration precise enough for modern inertial navigation techniques.

Another application of growing importance is a special type of electrical machine. Most conventional motors and generators depend for their operation on the force between magnetized iron parts; but in the homopolar machine the force

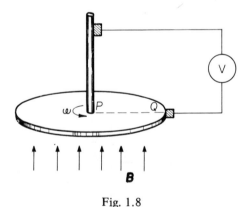

Fig. 1.8

is developed directly on a conductor moving in a magnetic field of constant intensity. Faraday in 1831 made the first generator using this principle, in which a circular disc rotates in a magnetic field (Fig. 1.8). Each element of the disc at a radial distance r from the axis is moving with velocity $u = \omega r$ perpendicular to the magnetic field. The induced electric field $u \times B$ is directed along the radius, so there will be an e.m.f. induced between the centre of the disc and the periphery. Integration along a path such as PQ gives

$$e = \int_P^Q u \times B \cdot ds = \int_0^a B\omega r \, dr = \tfrac{1}{2}B\omega a^2 \text{ volts,} \qquad (1.20)$$

where a is the radius of the disc in metres and ω the angular velocity in radians/second. The generated voltage is rather low for small machines at normal speeds, if the flux density is limited (as it usually is) by the saturation of iron to about 2 teslas. For example, if $a = 100$ mm, $\omega = 100$ rad/s (~ 1000 rev/min) and $B = 2$ T, then $e = 1$ V. Large machines have been built for special low-voltage heavy-current applications, and an important new development is the use of superconducting coils to generate very high magnetic fields. A motor rated at 2400 kW has recently been built using this principle (Advances in homopolar design [2]), and in large sizes the superconducting motor is expected to show economic advantages over conventional machines.

1.3 Electromagnetic induction

It was shown in section 1.2 that an e.m.f. is induced in a conductor when it moves in a magnetic field. An e.m.f. can also be induced in a stationary circuit by a time-varying magnetic field. If this magnetic field is produced by currents flowing in conductors or coils, the e.m.f. can be induced merely by changing the current; no motion of any part of the system is required. The effect is termed transformer induction, and it appears to be physically quite distinct from motional induction. Both effects are included in Faraday's law of electro-magnetic induction, which relates the induced e.m.f. in a circuit to the rate of change of the magnetic flux linking the circuit. Before discussing the application of this law it is necessary to establish the concepts of magnetic flux and flux linkage.

FLUX LINKAGE

If a circuit consists of a conductor in the form of a simple closed curve C, the magnetic flux Φ linking the circuit is defined by the surface integral

$$\Phi = \int_S B \cdot da \text{ webers,} \qquad (1.21)$$

where S is any surface spanning the boundary C of the circuit. If the magnetic field is uniform and the circuit has an area A perpendicular to the field, this reduces to the simple expression

$$\Phi = BA. \qquad (1.22)$$

The concept of flux linkage arises when it is desired to calculate the flux linking a multi-turn coil. It is possible in principle to devise a twisted surface resembling an Archimedian screw, bounded by the turns of the coil, and to evaluate the integral in equation 1.21 over this surface. But it is simpler to suppose that each turn links a certain amount of flux, so that the total flux linking the coil is the sum of the contributions from the individual turns. Thus if each turn links a flux Φ and the coil has n turns, then the total flux linking the coil, or flux linkage, is given by

$$\psi = n\Phi \text{ webers.} \tag{1.23}$$

If the magnetic field is uniform and parallel to the axis of the coil, and each turn has an area A, then

$$\psi = nBA. \tag{1.24}$$

Usually the field is not uniform and the flux through an individual turn will

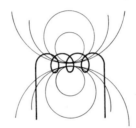

Fig. 1.9

depend on its position (Fig. 1.9). The total flux linking the coil is then given by the sum

$$\psi = \sum_{r=1}^{n} \Phi_r \text{ webers,} \tag{1.25}$$

and an average flux per turn may be defined by the relation

$$\Phi_{av} = \frac{\psi}{n}. \tag{1.26}$$

INDUCTANCE

If permanent magnets are excluded, the flux linking a coil will depend on (a) the current flowing in the coil, (b) currents flowing in any adjacent coils or conductors.

Self-inductance

With a single coil carrying a current i we have

$$\psi = f(i). \tag{1.27}$$

When there are no ferromagnetic materials present (i.e. the coil is air-cored) the relationship is linear, and

$$\psi = Li \tag{1.28}$$

where L is a constant known as the self-inductance of the coil. The unit of L is the henry when ψ is in webers and i in amperes. When the coil has an iron core the relationship between ψ and i is no longer linear, on account of the magnetic properties of iron. The form of equation 1.28 may still be used, but the coefficient L is no longer a constant; this can be made explicit by writing

$$\psi = iL(i). \tag{1.29}$$

In order to simplify the analysis it is often assumed that the inductance of an iron-cored coil is a constant; this assumption must be used with caution, for it can sometimes give completely erroneous results. This point will be discussed more fully in section 1.6.

Mutual inductance

With two coils the flux linkages are functions of the coil currents and the geometry of the system. Thus

$$\left. \begin{aligned} \psi_1 &= L_1 i_1 + M_{12} i_2, \\ \psi_2 &= L_2 i_2 + M_{21} i_1, \end{aligned} \right\} \tag{1.30}$$

where ψ_1 and ψ_2 are the flux linkages for the two coils, and i_1 and i_2 are the corresponding currents. The coefficients M_{12} and M_{21} are known as the mutual inductances. When the coils do not have iron cores, it may be shown (Carter[3]) that the mutual inductance coefficients are constant and equal, i.e.

$$M_{12} = M_{21} = M. \tag{1.31}$$

If ψ_{12} is the flux linking the first coil due to a current i_2 in the second, and ψ_{21} is the flux linking the second coil due to a current i_1 in the first, then the mutual inductance is given by

$$M = \frac{\psi_{12}}{i_2} = \frac{\psi_{21}}{i_1}. \tag{1.32}$$

This reciprocal property is particularly useful when the mutual inductance has to be measured or calculated, for one of the two alternative expressions in equation 1.32 is often easier to evaluate than the other.

FARADAY'S LAW

Faraday's law of electromagnetic induction states that the e.m.f. induced in a circuit is proportional to the rate of change of flux linkages. In SI units the constant of proportionality is unity, and

$$e = \pm \frac{d\psi}{dt} \text{ volts.} \qquad (1.33)$$

The question of the sign in equation 1.33 sometimes causes difficulty. Traditionally a negative sign is used in deference to Lenz's law, which states that any current produced by the e.m.f. tends to oppose the flux change. But this is

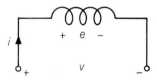

Fig. 1.10

inconsistent with the definition of inductance given in equation 1.28 and the usual circuit conventions shown in Fig. 1.10. If we take the positive sign in equation 1.33 and substitute for ψ from equation 1.28, then

$$e = +L\frac{di}{dt}. \qquad (1.34)$$

For a pure inductance, with no internal resistance, Kirchhoff's voltage law applied to the circuit gives

$$v - e = 0,$$

i.e.
$$v = e. \qquad (1.35)$$

Consequently the terminal voltage v is given by

$$v = +L\frac{di}{dt}, \qquad (1.36)$$

and this is the equation which defines the inductance element in circuit theory. The correct form of Faraday's law is therefore

$$e = +\frac{d\psi}{dt}, \qquad (1.37)$$

and Lenz's law may be used to resolve any uncertainty about the positive directions of e and ψ in the circuit.

CALCULATION OF THE INDUCED E.M.F.

Faraday's law relates the induced e.m.f. to the rate of change of flux linkages, regardless of the way in which the change occurs. The flux linkage of a circuit may be changed in several ways; the strength of the magnetic field may be altered, either by moving the circuit relative to the source of the field or by varying the currents which create the field; or the boundary of the circuit may be deformed while the magnetic field remains unchanged. The moving conductor in section 1.2 is an example of this last case; the circuit formed by the voltmeter, the fixed rails and the moving conductor steadily increases in area, and it will be seen that the rate of change of flux is equal to Blu in agreement with the previous calculation. In all cases the induced e.m.f. may be calculated by the direct application of Faraday's law, and this is the only satisfactory method when motional and transformer effects are both present.

Particular care is needed when calculating the e.m.f. in a moving conductor. It is tempting to use the flux cutting formula $e = Blu$ in all cases, but this can give incorrect results when parts of the magnetic structure move with the conductor. The derivation of the formula given in section 1.2 is for the particular case of a conductor whose motion does not affect the source of the magnetic field in any way, and its direct application is limited to that situation. More complex problems can be treated by expressing the total magnetic field as the sum of components from different parts of the magnetic structure, and taking the sum of Blu terms with the appropriate values of u (Binns [4]). But the direct application of Faraday's law is the safest procedure in this kind of problem. A good discussion of electromagnetic induction and some apparent paradoxes will be found in Carter [3].

INDUCED E.M.F. AND INDUCTANCE

The self- and mutual-inductance coefficients can often be changed by relative movement of parts of the system, and Faraday's law gives the correct value for the induced e.m.f. in these cases. For example, with the single coil shown in Fig. 1.10 the induced e.m.f. is given by

$$e = \frac{d\psi}{dt} = \frac{d}{dt}(Li) = L\frac{di}{dt} + i\frac{dL}{dt}. \tag{1.38}$$

Thus if the motion of a part of the system causes L to change, there will be an e.m.f. term additional to the normal e.m.f. of self induction. The voltage equation for the circuit should therefore be written as

$$v = Ri + \frac{d}{dt}(Li). \tag{1.39}$$

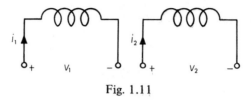

Fig. 1.11

Similarly, with the coupled coils shown in Fig. 1.11 the voltage equations are

$$\left.\begin{aligned} v_1 &= R_1 i_1 + \frac{d}{dt}(L_1 i_1 + M i_2), \\ v_2 &= R_2 i_2 + \frac{d}{dt}(L_2 i_2 + M i_1). \end{aligned}\right\} \tag{1.40}$$

When there is no motion of parts of the system the inductance coefficients are constant, and these equations reduce to the ordinary equations of coupled circuit theory.

1.4 Electromagnetic forces

In section 1.2 a method was given for calculating the force on a conductor in a magnetic field. In many practical devices, including rotating machines, magnetic forces act on the iron parts as well as on the conductors. These forces on the magnetized iron parts are often the dominant ones, and it is necessary to be able to calculate the total electromagnetic force acting on a structure made up of conductors and ferromagnetic materials. Two methods of calculation are given in this section. The first is the Maxwell stress method, which also provides a useful physical picture of the mechanism of force production. The second is an energy method, which complements the Maxwell stress method for purposes of calculation, but is less useful as a physical explanation.

THE MAXWELL STRESS CONCEPT

There is a sound scientific basis to the elementary idea that the magnetic lines of force are like rubber bands tending to draw pieces of iron together. The idea of lines or tubes of force was central to Faraday's conception of the magnetic field, but it was Maxwell who gave precise mathematical expression to this concept. Maxwell showed, as a deduction from the equations of the electromagnetic field, that magnetic forces could be considered to be transmitted through space (or a non-magnetic material) by the following system of stresses (Carter [3], Stratton [5]):

(a)   A tensile stress of magnitude $\frac{1}{2}BH$ newtons per square metre along the lines of force.

(b)   A compressive stress, also of magnitude $\frac{1}{2}BH$ newtons per square metre, at right angles to the lines of force.

Since $B = \mu_0 H$ in a non-magnetic medium, the stresses may also be written as $\mu_0 H^2 / 2$ or $B^2 / 2\mu_0$.

If the magnetic field is perpendicular to the surface of a body (Fig. 1.12), there will be a tensile stress of magnitude $B^2 / 2\mu_0$, also perpendicular to the surface, drawing the surface into the field. If the field is parallel to the surface (Fig. 1.13), there will be a compressive stress of magnitude $B^2 / 2\mu_0$ pushing the surface out of the field. In the general case, when the flux density B makes an angle θ with the normal n, the stress t makes an angle 2θ with n (Fig. 1.14).

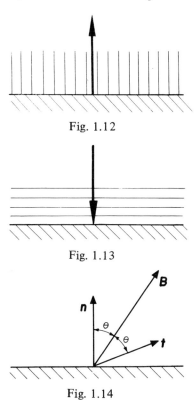

Fig. 1.12

Fig. 1.13

Fig. 1.14

The magnitude of t is still $B^2 / 2\mu_0$, and the three vectors n, B and t are coplanar. The force on an element of area δA is thus in the direction of t, and its magnitude is given by

$$\delta F = t \, \delta A = \frac{B^2}{2\mu_0} \, \delta A \text{ newtons.} \tag{1.41}$$

An interesting and at first sight surprising deduction is that the force will be parallel to the surface when the field is inclined at 45°.

It is sometimes useful to relate the components of stress to the components of flux density, which may be done by resolving the vectors in directions normal and tangential to the surface. Thus if B_n and B_s are respectively the normal and tangential components of B, the normal component of stress is given by

$$t_n = \frac{1}{2\mu_0}(B_n^2 - B_s^2),$$ (1.42)

and the tangential component is

$$t_s = \frac{B_n B_s}{\mu_0}.$$ (1.43)

CALCULATION OF THE FORCE FROM THE MAXWELL STRESS

The Maxwell stress concept gives an immediate qualitative picture of the way in which forces are exerted on the surface of an object in a magnetic field, and the method of calculation will be illustrated by two examples. Carpenter [6] has shown that the Maxwell stress method is generally superior to energy methods for this kind of problem.

Fig. 1.15 shows an electromagnet lifting an iron bar, and the problem is to calculate the force of attraction. The actual field pattern is quite complex, and it

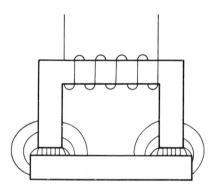

Fig. 1.15

would be very tedious first to solve the field equations with the boundary conditions imposed by the structure, and then to integrate the stress over the surface to find the force. A good approximation can be obtained from the following considerations. If the air gaps between the magnet poles and the bar are small, and the permeability of the magnetic material is high, the magnetic field in the gaps will be nearly uniform and much more intense than the fringing field outside the gaps. (This will be justified formally in section 1.6.) Since the Maxwell stress varies as B^2, the field outside the gaps will contribute very little to the total force, and it may be ignored. For the purpose of calculating the force, we may therefore replace the actual field distribution of Fig. 1.15 with the idealized distribution of Fig. 1.16, in which the field is uniform, confined to

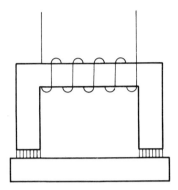

Fig. 1.16

the air gaps and normal to the iron surfaces. If the area of each pole face is A square metres, and the magnetic flux density in each gap is B teslas, the total force is

$$F = \frac{B^2}{2\mu_0} \cdot 2A = \frac{AB^2}{\mu_0} \text{ newtons.} \tag{1.44}$$

A method of calculating the flux density from the coil current and the dimensions of the magnet will be given in section 1.6.

Maxwell stress on a bounding surface

There are more complex problems which require another analytical device. It may be shown that the integral of the Maxwell stress over *any* surface surrounding a magnetic object gives the correct value for the total force on the object, even though the distribution of the force over this surface is quite different from the distribution over the surface of the object. As an example, consider the torque tending to rotate the short iron bar in Fig. 1.17 into alignment with the poles of the electromagnet. The permeability of the iron is assumed to be very high (ideally, infinite), so that the magnetic field is always normal to the iron surface. Evidently it is the stresses acting on the portions X and Y of the iron surface that are tending to rotate the bar; but the field here is particularly difficult to calculate. The field in the narrow air gap, on the other hand, is practically uniform and can be calculated by the methods of section 1.6. We therefore choose a surface such as *JKLM* (Fig. 1.18), and ignore the field beyond the corners *E, F, G* and *H*. The portions *EJ, FL, KG* and *MH* are perpendicular to the field; there will be tensile forces on these surfaces which cancel out in pairs.

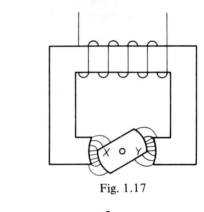

Fig. 1.17

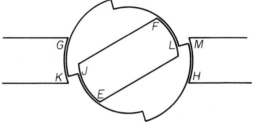

Fig. 1.18

Since the field is negligible beyond the corners, there is no force on the portions *FG* and *EH*. The portions *JK* and *LM* are both parallel to the field; there will be compressive stresses on these surfaces tending to rotate the bar in a clockwise direction. If *JK* = *LM* = *g* and the depth of the bar is *d*, the force on each surface is

$$F = \frac{B^2}{2\mu_0} \cdot gd \text{ newtons,}$$

and if *a* is the mean distance from the pivot to the air gap, the torque is

$$T = 2Fa = adgB^2/\mu_0 \text{ newton metres.} \tag{1.45}$$

Although the Maxwell stress concept has been introduced in terms of magnetized iron parts, it is not restricted to this situation. The electromagnetic force acting on *any* combination of iron parts and conductors may be found from the Maxwell stresses on a surface enclosing the bodies; the only restriction is that the surface should not pass through any magnetized parts. We shall see later that this provides a useful way of picturing the action of electrical machines.

ENERGY METHODS

Work must be done to establish a magnetic field, and the energy stored in the field is given by

$$W_m = \int \tfrac{1}{2} BH \, dv \text{ joules,} \tag{1.46}$$

where the integral is taken over the whole volume of the field.

Consider a system consisting of pieces of magnetic material together with coils or conductors carrying currents. In general there will be electromagnetic forces acting on the various parts of the system, and if any part is displaced the force will do work. Let there be a small displacement δx in some part of the system. If the component of force in the direction of the displacement is F_x, the work done will be

$$\delta W = F_x \, \delta x. \tag{1.47}$$

During this displacement there may be an increase δW_m in the stored magnetic energy, and if voltages are induced in any of the coils the electrical sources will have to supply an amount of energy δW_e. We thus have

$$\text{energy supplied} = \frac{\text{increase in}}{\text{stored energy}} + \text{work done}$$

i.e.
$$\delta W_e = \delta W_m + \delta W$$
$$= \delta W_m + F_x \, \delta x. \tag{1.48}$$

If the currents in the coils are adjusted continuously during the displacement so that there is no change in the flux linkages, there will be no induced voltages; consequently the energy supplied, δW_e, will be zero. The work done by the force must come from the energy stored in the field, and

$$F_x \, \delta x = -\delta W_m,$$

i.e.
$$F_x = - \frac{\partial W_m}{\partial x} \bigg|_{\text{constant flux.}} \tag{1.49}$$

When there is a linear relationship between flux and current another expression may be obtained. If the currents in all the coils are held constant during the displacement, it may be shown (Carter [3]) that the energy δW_e supplied by the sources is equally divided between the mechanical work $F_x \delta x$ and the increase in stored energy δW_m. Thus,

$$F_x \delta x = \delta W_m,$$

i.e.
$$F_x = + \frac{\partial W_m}{\partial x} \Bigg|_{\text{constant current.}} \qquad (1.50)$$

The force will be in newtons when the displacement is in metres and the field energy is in joules. Similar equations hold for rotational motion if F_x is replaced by the torque T_θ (in newton metres) and x is replaced by the angular displacement θ (in radians).

Calculation of the force on an iron part

As an example of the use of these expressions, consider once again the electromagnet shown in Fig. 1.16. If x is the displacement of the bar from the poles, the field energy is given by

$$W_m = \int_{\text{air gap}} \tfrac{1}{2} BH \, dv \;+\; \int_{\text{core}} \tfrac{1}{2} BH \, dv$$

$$\qquad (1.51)$$

$$= \frac{B^2 Ax}{\mu_0} + \int_{\text{core}} \tfrac{1}{2} BH \, dv \text{ joules.}$$

If ψ is constant, B will be constant and the energy stored in the core will also be constant. Therefore

$$F_x = - \frac{\partial W_m}{\partial x} \Bigg|_{\text{constant flux}} = - \frac{B^2 A}{\mu_0} \text{ newtons.} \qquad (1.52)$$

This is numerically the same as equation 1.44 obtained from the Maxwell stresses, and the negative sign shows that the force on the bar is in the direction of decreasing x, i.e. upwards. In this case the Maxwell stress method is obviously simpler.

ENERGY AND INDUCTANCE

With a single coil carrying a current i, the energy stored in the magnetic field is given by

$$W_m = \tfrac{1}{2} Li^2 . \qquad (1.53)$$

If the motion of a part of the system causes a change in the inductance, then equation 1.50 gives

$$F_x = \frac{\partial W_m}{\partial x}\Bigg|_{\text{constant current}} = \tfrac{1}{2}i^2\,\frac{\partial L}{\partial x}\,. \qquad (1.54)$$

This is an important and useful result, for it shows that the mechanical force can be expressed in terms of the variation in the inductance coefficient L, a quantity which can be measured electrically.

With a pair of mutually coupled coils carrying currents i_1 and i_2, the stored magnetic energy is given by

$$W_m = \tfrac{1}{2}L_1 i_1^2 + \tfrac{1}{2}L_2 i_2^2 + M i_1 i_2. \qquad (1.55)$$

The force acting on a part of the system is then given by

$$F_x = \tfrac{1}{2}i_1^2\,\frac{\partial L_1}{\partial x} + \tfrac{1}{2}i_2^2\,\frac{\partial L_2}{\partial x} + i_1 i_2\,\frac{\partial M}{\partial x}\,. \qquad (1.56)$$

Similar expressions hold for torque in terms of angular displacement.

Calculation of the torque on an air-cored coil

As an example of the application of this force expression, consider the element of an electrodynamic wattmeter, shown diagrammatically in Fig. 1.19. A small moving coil is mounted on pivots mid-way between two fixed coils. The fixed coils are connected in series, and are separated by a distance equal to their

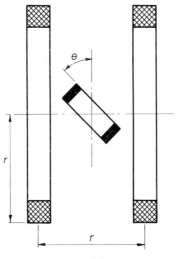

Fig. 1.19

radius; this is a Helmholtz pair, which produces in the vicinity of the moving coil a nearly uniform magnetic field parallel to the common axis of the fixed coils. If each fixed coil has N_1 turns and carries a current i_1 amperes, the flux density is

$$B = \frac{8\mu_0 N_1 i_1}{5\sqrt{5}r} \text{ teslas,} \qquad (1.57)$$

where r is the radius in metres.

If the moving coil has N_2 turns and an area A square metres, the flux linking it is

$$\psi_{21} = N_2 AB \cos\theta \text{ webers,} \qquad (1.58)$$

and the mutual inductance between the fixed and moving coils is

$$M = \frac{\psi_{21}}{i_1} = \frac{8\mu_0}{5\sqrt{5}} \cdot \frac{N_1 N_2 A}{r} \cos\theta \text{ henrys.} \qquad (1.59)$$

Since the self inductances are independent of θ, the torque on the moving coil is

$$T = i_1 i_2 \frac{\partial M}{\partial \theta} = -\frac{8\mu_0}{5\sqrt{5}} \cdot \frac{N_1 N_2 A}{r} i_1 i_2 \sin\theta \text{ newton metres.} \qquad (1.60)$$

This is an instance of an energy method giving a straightforward calculation of the torque, whereas the Maxwell stress would be difficult to evaluate.

1.5 Magnetic materials

In free space the magnetic flux density B is related to the magnetizing force H by the expression

$$B = \mu_0 H, \qquad (1.61)$$

where μ_0 is the primary magnetic constant (with a value of $4\pi \times 10^{-7}$ H/m). In a material medium, the relation becomes

$$B = \mu_0 \mu_r H, \qquad (1.62)$$

where μ_r is a dimensionless number known as the relative permeability of the material. For most engineering purposes materials may be divided into two groups: the ferromagnetic materials, typified by iron, for which the relative permeability μ_r is large and variable; and all other materials, for which μ_r is practically equal to unity. There is normally no need to consider the small deviations from unity which characterize paramagnetic and diamagnetic behaviour. Only a brief introduction to the properties of ferromagnetic materials will be given in this section; further information is available in standard texts such as Brailsford [7].

A typical ferromagnetic material is silicon steel, which is widely used for the cores of transformers and rotating machines. When such a material is magnetized by slowly increasing the applied magnetizing force H, the resulting flux density B follows a curve of the form shown in Fig. 1.20. This is known as the magnetization curve for the material, and the corresponding variation of μ_r with B is

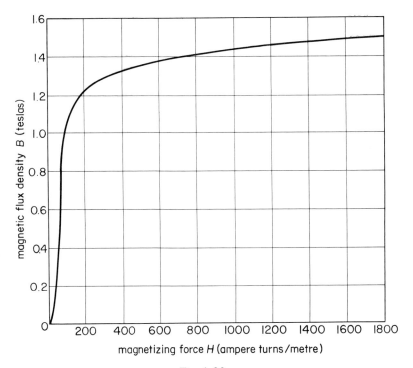

Fig. 1.20

shown in Fig. 1.21. If the magnetizing force is gradually reduced to zero, the flux density does not follow the same curve; and if the magnetizing force slowly alternates between positive and negative values, the relationship between B and H takes the form of a hysteresis loop as shown in Fig. 1.22. When the amplitude of the alternating magnetizing force is changed, a new hysteresis loop will be formed; and the locus of the tips of these loops is the magnetization curve shown in Fig. 1.20.

The part of the magnetization curve where the slope begins to change rapidly is termed the knee. Below the knee it is often possible to use a linear approximation to the actual characteristic, with a corresponding constant value for the relative permeability. But the onset of saturation above the knee marks a

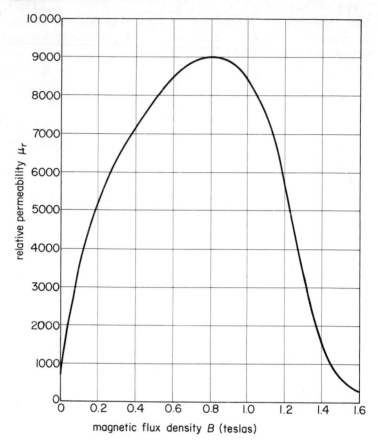

Fig. 1.21

dramatic change in the properties of the material, which must be recognized in the design and analysis of magnetic structures.

HYSTERESIS LOSS

When a magnetic material is taken through a cycle of magnetization, energy is dissipated in the material in the form of heat. This is known as the hysteresis loss, and it may be shown (Carter [3]) that the energy loss per unit volume for each cycle of magnetization is equal to the area of the hysteresis loop. The area of the loop will depend on the nature of the material and the value of B_{max} (Fig. 1.22), and an approximate empirical relationship discovered by Steinmetz is

$$w_h = \lambda_h B_{max}^n \text{ joules/metre}^3.$$ (1.63)

In this expression w_h is the loss per unit volume for each cycle of magnetization; the index n has a value of about 1.7 for many materials; and the coefficient λ_h is a property of the material, with typical values of 500 for 4% silicon steel and 3 000 for cast iron.

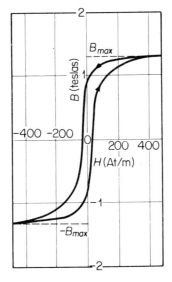

Fig. 1.22

When the material is subjected to an alternating magnetic field of constant amplitude there will be a constant energy loss per cycle, and the power absorbed is therefore proportional to the frequency. Assuming the Steinmetz law, we have the following expression for the hysteresis loss per unit volume:

$$p_h = \lambda_h \, B_{max}^{1.7} f \text{ watts/metre}^3, \qquad (1.64)$$

where f is the frequency in hertz.

EDDY CURRENT LOSS

If a closed loop of wire is placed in an alternating magnetic field, the induced e.m.f. will circulate a current round the loop. A solid block of metal will likewise have circulating currents induced in it by an alternating field, as shown in Fig. 1.23. These are termed eddy currents, and they are a source of energy loss in the metal. Eddy current losses occur whenever conducting material is placed in a changing magnetic field; the magnitude of the loss is dependent on the properties of the material, its dimensions and the frequency of the alternating field.

Magnetic structures carrying alternating magnetic flux are usually made from a stack of thin plates or laminations, separated from one another by a layer of insulation (Fig. 1.24). This construction breaks up the eddy current paths, with a consequent reduction in the loss; qualitatively, the effect may be explained as follows. With solid metal (Fig. 1.23) the currents would flow in approximately

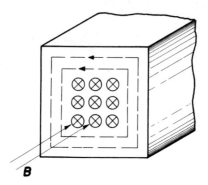

Fig. 1.23

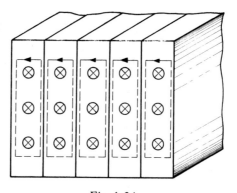

Fig. 1.24

square paths; these paths enclose a large area for a given perimeter, and the induced e.m.f. is high for a path of given resistance. When the metal is divided into laminations (Fig. 1.24), the current paths are long narrow rectangles; the area enclosed by a given perimeter is much smaller, and the induced e.m.f. is smaller, giving lower currents and reduced losses. An approximate analysis (Seely [8]) shows that in plates of thickness t (where t is much smaller than the width or length) the eddy current loss per unit volume is given by

$$p_e = \frac{\pi^2 B_{max}^2 f^2 t^2}{6\rho} \text{ watts/metre}^3, \qquad (1.65)$$

where the flux density is an alternating quantity of the form

$$B = B_{max} \sin 2\pi ft, \tag{1.66}$$

and ρ is the resistivity of the material. Thus if the lamination thickness is reduced by a factor k, the loss is reduced by a factor k^2. As might be expected, the loss varies inversely with the resistivity ρ. The addition of 3–4% of silicon to iron increases the resistivity by about four times, as well as reducing the hysteresis loss; this is the main reason for the widespread use of silicon steel in electrical machines. The thickness of the laminations is usually about 0.5 mm, which ensures that the eddy current loss will be less than the hysteresis loss at a frequency of 50 Hz.

SKIN EFFECT

The eddy currents in a bar such as the one shown in Fig. 1.23 will produce a magnetic field within the bar which, by Lenz's law, will oppose the applied field. Thus the magnetic flux density will fall from a value B_0 at the surface to some lower value in the interior. The effect depends on the properties of the material, the frequency of the alternating field and the dimensions of the bar. It is possible for the magnitude of the flux density to fall very rapidly in the interior of the bar, so that most of the flux is confined to a thin layer or skin near the surface. The phenomenon is termed skin effect, and it implies very inefficient use of the magnetic material (quite apart from any eddy current losses). A similar effect occurs in conductors carrying alternating current, where the current density falls from some value J_0 at the surface to a lower value in the interior.

The variation of flux density with distance may be calculated by solving the electromagnetic field equations. When skin effect is well developed (i.e. the flux density decays rapidly) the solution is independent of the geometry of the bar, and the magnitude of the flux density is given by

$$B = B_0 e^{-x/\delta}, \tag{1.67}$$

where x is the distance into the material from the surface. The quantity δ is given by

$$\delta = \sqrt{\frac{2\rho}{\mu\omega}}, \tag{1.68}$$

where ω is the angular frequency of the alternating field, ρ is the resistivity of the material and $\mu = \mu_0 \mu_r$ is its permeability (assumed constant). At a depth δ the magnitude of B is $1/e$ times the value at the surface, and δ is known as the depth of penetration or the skin depth. A full discussion of skin effect in conductors and magnetic materials is given in Carter [3].

The phenomenon of skin effect gives a second reason for using laminated magnetic circuits. If the thickness of a plate is much more than twice the depth of penetration δ, the central region will carry very little flux. The material will be fully utilized if it is divided into laminations less than δ in thickness, for the flux density will then be fairly uniform across the lamination. The depth of penetration in silicon steel is about 1 mm at a frequency of 50 Hz, so the usual lamination thickness of 0.5 mm ensures that skin effect will not be significant.

1.6 The magnetic circuit

In the study of electromagnetic devices, it is often necessary to determine the magnetic field from a knowledge of the structure of the device and the magnitudes of the currents flowing in coils or other conductors. An exact determination involves the solution of the partial differential equations of the electromagnetic field, a problem which is rendered extremely difficult by the non-linear properties of magnetic materials. The magnetic circuit concept provides an approximate method of solution which is good enough for most design purposes and gives some immediate physical insight into the behaviour of magnetic structures.

THE MAGNETIC CIRCUIT CONCEPT

Fig. 1.25 shows a closed iron core magnetized by a coil carrying a current. If the relative permeability of the iron is high, most of the magnetic flux will be confined to the iron. The flux Φ through any cross-section of the core will then

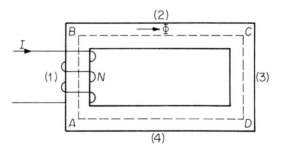

Fig. 1.25

be substantially the same; this follows from the fact that the flux of B out of any closed surface is zero; if there is no flux out of the sides of the iron, the flux entering a section must be equal to the flux leaving it. We thus have an analogy with a simple electric circuit (Fig. 1.26), in which the current i (which is the flux of the current density J) is the same for all cross-sections of the conductor.

Corresponding to Kirchhoff's current law $\Sigma i = 0$ we have the flux law $\Sigma \Phi = 0$, and the structure of Fig. 1.25 is known as a *magnetic circuit*. We shall see that there is a very close analogy between the analysis of the electric and magnetic circuits.

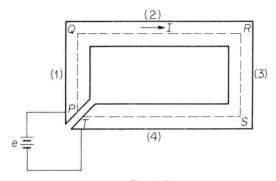

Fig. 1.26

Apply Kirchhoff's voltage law to the electric circuit of Fig. 1.26, following the path *PQRST*:

$$e = -\Sigma v \tag{1.69}$$
$$= -(v_{PQ} + v_{QR} + v_{RS} + v_{ST})$$

$$= \int_P^Q E \cdot ds + \int_Q^R E \cdot ds + \int_R^S E \cdot ds + \int_S^T E \cdot ds. \tag{1.70}$$

The battery voltage e is the applied *electromotive force* (e.m.f.) in the circuit.

Now apply Ampere's circuital law to the magnetic circuit of Fig. 1.25, following the path *ABCDA*:

$$NI = \oint H \cdot ds$$

$$= \int_A^B H \cdot ds + \int_B^C H \cdot ds + \int_C^D H \cdot ds + \int_D^A H \cdot ds. \tag{1.71}$$

The similarity between equations 1.70 and 1.71 is obvious. Just as the quantity

$$v_{PQ} = -\int_P^Q E \cdot ds$$

is the electric potential difference between P and Q, the quantity

$$U_{AB} = -\int_A^B H \cdot ds$$

is the magnetic potential difference between A and B. If we now put $NI = F*$, equation 1.71 becomes

$$F = \oint H \cdot ds$$
$$= -(U_{AB} + U_{BC} + U_{CD} + U_{DA}) \tag{1.72}$$
$$= -\Sigma U,$$

and equation 1.72 is the magnetic counterpart of Kirchhoff's voltage law (equation 1.69). By analogy with the electromotive force, the quantity $F = NI$ is known as the *magnetomotive force* (m.m.f.), measured in ampere turns (At). It follows that the units of magnetic potential, U, are also ampere turns.

ELECTRIC AND MAGNETIC CIRCUIT ANALOGIES

The analogy can be taken a stage further. In Fig. 1.26, suppose that limb (1) has a length l_1, a cross-sectional area A_1, and a constant conductivity σ_1. Since the electric field is practically uniform,

$$\int_P^Q E \cdot ds = E_1 l_1$$
$$= \frac{J_1 l_1}{\sigma_1}$$
$$= \frac{i l_1}{\sigma_1 A_1}$$
$$= i R_1,$$

where R_1 is the *resistance* of limb (1) given by

$$R_1 = \frac{l_1}{\sigma_1 A_1} \text{ ohms.} \tag{1.73}$$

Thus equation 1.70 becomes

$$e = i(R_1 + R_2 + R_3 + R_4). \tag{1.74}$$

In Fig. 1.25, suppose likewise that limb (1) has a length l_1, a cross-sectional area A_1, and a *constant* permeability μ_1 (where $\mu = \mu_0 \mu_r$). Since the magnetic

---

* Note that the symbol F in this context does *not* represent mechanical force.

field is practically uniform,

$$\int_A^B \mathbf{H} \cdot ds = H_1 l_1$$

$$= \frac{B_1 l_1}{\mu_1}$$

$$= \frac{\Phi l_1}{\mu_1 A_1}$$

$$= \Phi S_1,$$

where S_1 is known as the *reluctance* of limb (1) and is given by

$$S_1 = \frac{l_1}{\mu_1 A_1} \text{ ampere turns/weber.} \qquad (1.75)$$

Thus equation 1.71 becomes

$$NI = F = \Phi(S_1 + S_2 + S_3 + S_4), \qquad (1.76)$$

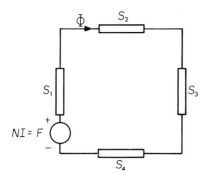

Fig. 1.27

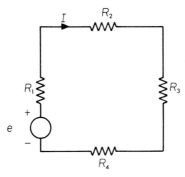

Fig. 1.28

The similarity between equations 1.73 and 1.75, and between equations 1.74 and 1.76 should be noted. It follows that the electric and magnetic circuits may be represented by the *circuit diagrams* of Figs. 1.27 and 1.28. Table 1.1 summarizes the analogy between electric and magnetic circuits.

Table 1.1

Electric and magnetic circuit analogies

Electric circuit	Magnetic circuit
Electromotive force e	Magnetomotive force F
Electric current i	Magnetic flux Φ
Electric potential difference v	Magnetic potential difference U
Resistance $R = v/i = l/\sigma A$	Reluctance $S = U/\Phi = l/\mu A$
Current law $\Sigma i = 0$	Flux law $\Sigma \Phi = 0$
Voltage law $\Sigma(e + v) = 0$	Circuital law $\Sigma(F + U) = 0$

SIMPLE MAGNETIC CIRCUITS WITH AIR GAPS

An example will illustrate the methods of analysing a simple magnetic circuit. Fig. 1.29 shows an electromagnet consisting of an iron core with an air gap. The coil carries a current of 1 A, and we wish to find the number of turns required to set up a flux density of 1.2 T in the air gap.

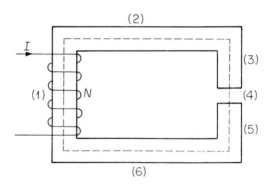

Fig. 1.29

A simple method is to assume a constant value for the permeability of the core, and then to calculate the reluctances of the various parts of the magnetic circuit. If the core material is silicon steel, Fig. 1.21 shows that the relative permeability exceeds 3 000 for a wide range of flux densities below the knee of the magnetization curve. Table 1.2 gives the dimensions of the magnetic circuit and the calculated results with a value of 3 000 for the relative permeability of the core.

Table 1.2

Magnetic circuit calculation using constant permeability

Section of circuit	(1)	(2)	(3)	(4)	(5)	(6)
Material	iron	iron	iron	air	iron	iron
Relative permeability μ_r	3 000	3 000	3 000	1	3 000	3 000
Length l (mm)	150	250	74	2	74	250
Area A (mm^2)	80	125	100	100	100	125
Reluctance $S = l/\mu_0\mu_r A$						
(MAt/Wb)	0.50	0.53	0.20	15.9	0.20	0.53

Reluctance of iron path = 1.96×10^6 At/Wb
Reluctance of air gap = 15.9×10^6 At/Wb
Total reluctance   = 17.9×10^6 At/Wb

The flux in the core is given by

$$\Phi = BA = 1.2 \times 100 \times 10^{-6} \text{ Wb}$$

$$= 120 \,\mu\text{Wb},$$

and the m.m.f. is therefore

$$F = \Phi S = 120 \times 10^{-6} \times 17.9 \times 10^6 \text{ At}$$

$$= 2\,170 \text{ At.}$$

Since the coil current is 1 A the required number of turns is 2 170.

A second and more accurate method of analysing the magnetic circuit is to work directly from Ampere's law in the form

$$F = NI = \oint H.ds = \Sigma Hl. \qquad (1.77)$$

The specified air-gap flux density and area define the flux Φ, and the flux density in any other part of the circuit is given by the relation $B = \Phi/A$. The corresponding value of H may be found from the magnetization curve for the material used in that part of the circuit, and the sum of the Hl terms may then be computed. Table 1.3 shows the calculation; the value of H in the air gap is obtained from the relation $B = \mu_0 H$; the magnetization curve of Fig. 1.20 is used for the iron parts; and the flux is given by $\Phi = BA = 120 \,\mu\text{Wb}$. The number of turns required on the coil is therefore 2 250, and the total reluctance given by this calculation is

$$S = F/\Phi = 18.7 \times 10^6 \text{ At/Wb.}$$

Table 1.3

Magnetic circuit calculation using the magnetization curve

Section of circuit	(1)	(2)	(3)	(4)	(5)	(6)
Material	iron	iron	iron	air	iron	iron
Area A (mm^2)	80	125	100	100	100	125
Flux density $B = \Phi/A$(T)	1.5	0.96	1.2	1.2	1.2	0.96
Magnetizing force H (AT/m)	1 800	90	170	954×10^3	170	90
Length l (mm)	150	250	74	2	74	250
Potential drop Hl (At)	270	22.5	12.6	1910	12.6	22.5

Potential drop in iron   =   340 At
Potential drop in air gap = 1 910 At
Total potential drop = F= 2 250 At

This is larger than the previous result of 17.9×10^6 At/Wb, and an examination of Table 1.3 shows that section (1) of the magnetic circuit has been driven into saturation, with a flux density of 1.5 T. This results in an excessive potential drop in the section, with a corresponding increase in the total reluctance. If the air-gap flux density is reduced to 1.0 T, section (1) will come out of saturation with a flux density of 1.25 T. The whole iron path is then unsaturated, and a similar calculation gives a value of 16.7×10^6 At/Wb for the total reluctance. This is smaller than the value of 17.9×10^6 At/Wb calculated on the assumption of a constant permeability of 3 000, because the relative permeability of most parts of the circuit is now greater than 3 000, but the error is not large.

LINEARITY

In the example just considered the iron path is nearly 400 times the length of the air gap, but it only contributes about 10% to the total reluctance of the magnetic circuit. Thus when the iron is unsaturated quite a small air gap will have a dominant effect, and any changes in the magnetic condition of the core will cause very little change in the total reluctance. Since $NI = \Phi S$, it follows that the flux will be proportional to the current if the total reluctance is constant, and this is the basis of the assumption of linearity made earlier in the chapter. There are two situations in which the relationship between flux and current is not even approximately linear. If there are no air gaps in the magnetic circuit, there will be no constant reluctance term to swamp the variable reluctance of the iron, and the relationship between flux and current will be determined by the magnetization curve of the material. If there are air gaps but the iron is driven into saturation, the relative permeability of the iron will be low (see Fig. 1.21) and its reluctance may be comparable with the reluctance of the gaps; the non-linear iron characteristic will again make its presence felt. The reader is invited to explore these effects in problem 1.8 at the end of the chapter.

FRINGING AND LEAKAGE

In the analysis of a magnetic circuit with an air gap, two assumptions have been made in order to simplify the calculation:

(1)   Flux passes straight across the air gap, without spreading into the surrounding air.

(2)   There is no leakage of flux from the iron path into the surrounding air.

In practice, there is some spreading of the air-gap flux (known as *fringing*), and leakage cannot be neglected when the air gap is large (i.e. when its length is not negligible in comparison with the other air spaces between the iron parts). Both of these effects are illustrated in Fig. 1.30. It is possible to introduce fringing

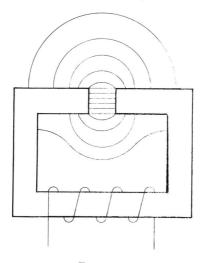

Fig. 1.30

and leakage coefficients to take account of these effects, thereby extending the magnetic circuit theory to handle quite complex problems without having to resort to a full field analysis (M.I.T. Staff [9]).

MAGNETIC CIRCUITS WITH PARALLEL PATHS

Fig. 1.31 shows a magnetic circuit with parallel paths. It is only possible to analyse this circuit easily if we assume that the permeability is constant, as well as neglecting leakage and fringing. The circuit can then be represented by the diagram of Fig. 1.32, where S_a and S_i are the *constant* reluctances of the air and iron paths, and the circuital law is used in the form $F = -\Sigma U = \Sigma \Phi S$. Applying

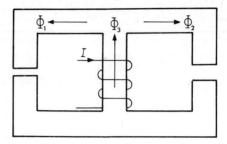

Fig. 1.31

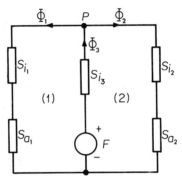

Fig. 1.32

the flux law to the junction P gives $\Phi_3 = \Phi_1 + \Phi_2$, and applying the circuital law to meshes (1) and (2) gives

$$F = S_{i_3}(\Phi_1 + \Phi_2) + (S_{a_1} + S_{i_1})\Phi_1, \qquad (1.78)$$

$$F = S_{i_3}(\Phi_1 + \Phi_2) + (S_{a_2} + S_{i_2})\Phi_2. \qquad (1.79)$$

Note that this is exactly analogous to applying Kirchhoff's voltage law in the form $e = \Sigma iR$ to the two meshes of a similar electric circuit. More complex magnetic circuits may be handled in a similar way, making full use of any relevant electric circuit theorems.

MAGNETIC CIRCUITS: CONCLUDING REMARKS

The circuit analogy is a useful device for deducing the properties of magnetic structures from the more familiar properties of electric circuits. An e.m.f. drives a current through an electric circuit against the resistance; an m.m.f. drives a flux through a magnetic circuit against the reluctance, and the greater the reluctance the greater the number of ampere turns (m.m.f.) required to establish the flux. Just as current takes the path of least resistance, flux takes the path of least reluctance. In magnetic circuits with small air gaps the flux will be concentrated in the low-reluctance region of the gap, and the fringing field beyond the gap will fall away rapidly as the length of the air path increases. Fig. 1.33 shows a type of structure commonly found in electrical machines, and it is immediately evident that the flux will be concentrated in the teeth, leaving a relatively weak field in the slots. An approximate calculation follows from the

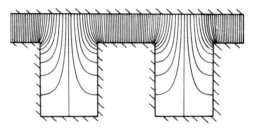

Fig. 1.33

electric circuit rule for current division in parallel branches: the flux will divide in the ratio of the permeances, where permeance (symbol Λ), the reciprocal of reluctance, is analogous to the conductance of an electrical circuit.

Problems

1.1 A flat copper plate is held in a vertical plane and then allowed to fall through the gap between the poles of a magnet. The plate is much wider than the poles, and any induced currents which flow in the part of the plate between the poles are assumed to find return paths of negligible resistance in the rest of the plate. The magnet poles are square, and the magnetic field may be assumed to be uniform and confined to the area of the poles. Show that the plate will experience a retarding force proportional to its velocity, and calculate the magnitude of the force when the velocity is 10 m/s. The thickness of the plate is 5 mm; the resistivity of copper is 1.68×10^{-8} Ωm; the magnet poles are 100 mm square; and the magnetic flux density is 1.0 T.

1.2 In the Faraday disc machine (p. 8) there will be a torque on the disc when current flows between the centre and the periphery. By considering the torque on an elementary annular ring of the disc, show that the total torque is given by the expression

$$T = \tfrac{1}{2} Bia^2 \text{ newton metres}$$

where B is the magnetic flux density in teslas, i is the current in amperes and a

is the radius of the disc in metres. Hence show that the mechanical power output from the disc is equal to the electrical power input when the machine runs as a motor. The resistance of the disc may be ignored.

1.3 Fig. 1.34 shows an extract from a patent specification (British Patent [10]) for an improved type of d.c. generator. The rotor R is magnetized by the field coil F, and thus forms a rotating magnet. The inventors claim that the rotating

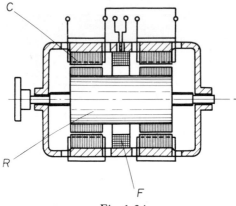

Fig. 1.34

field of this magnet will induce an e.m.f. in a stationary conductor such as C, which forms one side of a rectangular coil. Several coils may be connected in series to increase the output voltage of the machine.

Some engineers have expressed doubts as to whether the machine will actually work. Settle the question by the application of Faraday's law.

1.4 Show that there can be no magnetic field outside a coaxial cable carrying a current. Hence show that an alternating magnetic field due to currents in other conductors will induce no voltage in the cable circuit.

1.5 A steel ring is uniformly wound with a coil of 1 000 turns, and a flux density of 1.5 T is produced in the ring by a coil current of 3 A. If the ring has a mean diameter of 0.2 m and a cross-sectional area of 0.001 m^2, calculate the inductance of the coil.

The ring is divided into two semi-circular parts by making cuts in the iron. The width of each cut is 2 mm, and the current in the coil is increased to maintain the original flux density of 1.5 T. Calculate (a) the new value of the coil current; (b) the new coil inductance; (c) the force of attraction between the two halves of the ring.

1.6 The device shown in Fig. 1.17 will function as an elementary form of a.c. motor when the coil carries a current of the form $i = I_m \cos \omega t$. The inductance of the coil depends on the rotor position and may be assumed to follow the law $L = L_1 + L_2 \cos 2\theta$, where θ is the angular position of the rotor. If the rotor revolves with a steady angular velocity ω_r, so that $\theta = \omega_r t + \phi$, show that the torque on the rotor is given by the expression

$$T = -\tfrac{1}{2} L_2 I_m^2 (1 + \cos 2\omega t) \sin(2\omega_r t + 2\phi).$$

Show that the torque will have an average value of zero unless $\omega_r = \omega$, and obtain an expression for the average value when this condition holds.

1.7 The following figures were obtained for the power loss in the core of a transformer at different frequencies, with the maximum value of the flux density held constant.

frequency (Hz)	35	40	45	50	55	60	65	70
power loss (W)	46	54	62	70	78	87	96	105

If the loss is made up of hysteresis and eddy current components, show that the total loss should be related to the frequency by an expression of the form

$$P = Af + Bf^2$$

where A and B are constants. By plotting a graph of the energy loss per cycle against frequency determine the constants A and B for the transformer, and hence find the values of the hysteresis and eddy current components of the core loss at a frequency of 50 Hz.

1.8 Assume that the ring in problem 1.5 is made from silicon steel, with the magnetization curve shown in Fig. 1.20. Calculate the values of coil current required to give a number of values of flux density ranging from 0 to 1.5 T, (a) for a solid ring, (b) for a ring with a single air gap of 1 mm. Hence plot graphs of flux against current and inductance against current for the two cases.

References

[1] DUNSHEATH, P. (1962). *A history of electrical engineering*. Faber, London.
[2] Advances in homopolar machine design. *Electrical Review*, 4 April 1969, **184**, 488–489.
[3] CARTER, G. W. (1967). *The electromagnetic field in its engineering aspects*, 2nd edition. Longmans, London.
[4] BINNS, K. J. (1963). Flux cutting or flux linking. *J. IEE*, **9**, 259.
[5] STRATTON, J. A. (1941). *Electromagnetic theory*. McGraw-Hill, New York.
[6] CARPENTER, C. J. (1960). Surface integral methods of calculating forces on magnetized iron parts. *Proc. IEE*, **107C**, 19–28.

[7] BRAILSFORD, F. (1966). *Physical principles of magnetism*. Van Nostrand, London.

[8] SEELY, S. (1958). *Introduction to electromagnetic fields*. McGraw-Hill, New York.

[9] M.I.T. STAFF (1943). *Magnetic circuits and transformers*. Wiley, New York.

[10] BRITISH PATENT 917 263 (1963).

CHAPTER 2

D.C. Machines

Historically, d.c. machines were the first to be developed because the only available electrical power source was the d.c. voltaic cell. The advantages of alternating current were later recognized, and the invention of the induction motor was an important factor in securing acceptance of the alternating current system. The two main types of a.c. machine (synchronous and induction machines) are structurally simpler than d.c. machines; but the theory of a.c. machines is inherently more complex than the theory of d.c. machines, and we therefore adopt the historical order in developing the principles.

The homopolar machine mentioned in section 1.2 is a d.c. machine; in fact it is a 'pure' d.c. machine, for we shall see that the conventional machine generates an alternating voltage which is rectified mechanically by the commutator. Before the advent of cryogenics, the homopolar machine was only suitable for certain low-voltage heavy-current applications, and the development of d.c. machines followed a different path. If a coil rotates in a magnetic field (Fig. 2.1), the flux linking the coil will be an alternating quantity, and an alternating e.m.f. will be induced in the coil. In Fig. 2.1 connection is made to the coil by brushes

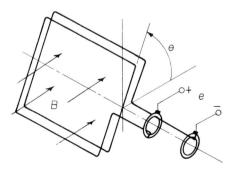

Fig. 2.1

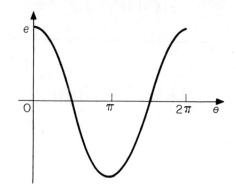

Fig. 2.2

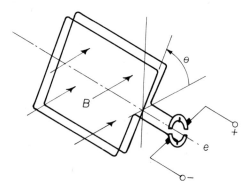

Fig. 2.3

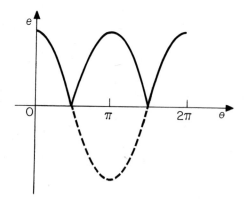

Fig. 2.4

bearing on sliprings; an alternating voltage is developed at the terminals (Fig. 2.2) which reverses with every half revolution of the coil. Suppose that instead of being connected to separate rings, the two ends of the coil are connected to the two parts of a divided ring (Fig. 2.3). This arrangement reverses the connections to the coil every half revolution, so the terminal voltage is now unidirectional (Fig. 2.4); this is a primitive d.c. machine, and the divided ring is termed a commutator. The advantage of this arrangement over the homopolar machine is that the generated voltage can be made as large as we please by winding enough turns on the coil.

2.1 Fundamental principles

In most d.c. machines the magnetic field is provided by an electromagnet, and to minimize the ampere turns required for the field coil the air gap is made as small as possible. The active (armature) coils are therefore wound on an iron ring or cylinder, and Fig. 2.5 shows a cross-section through a simple machine

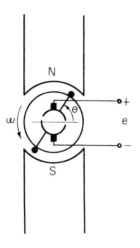

Fig. 2.5

with a single-turn armature coil wound on the surface of a smooth iron cylinder. (The field coils are not shown in this figure.) Practical machines differ from this model in having many more coils distributed round the periphery of the armature, and the conductors are usually embedded in slots instead of being placed on the surface of the cylinder. We first analyse the simple model, and then show how the results may be generalized for the practical machine.

VOLTAGE AND TORQUE EQUATIONS FOR A SIMPLE MODEL

Take the model shown in Fig. 2.5, and assume that the magnetic field in the air gap is purely radial, and uniform along the length of the armature. The flux density B will, of course, vary with the angular position round the air gap, as shown in Fig. 2.6. Only the coil sides in the air gap will be considered; assume

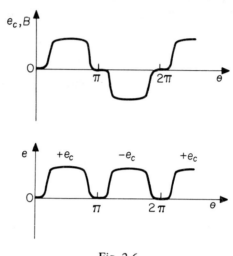

Fig. 2.6

that the magnetic field outside the gap is negligible, so that the rest of the coil contributes nothing to the e.m.f. or torque.

As the armature rotates, the coil sides move in a magnetic field; an e.m.f. will be generated in each conductor, given by equation 1.5:

$$e = Blu \text{ volts.}$$

The e.m.f. developed in the whole coil is therefore

$$e_c = 2Blu = 2Blr\omega \text{ volts,} \tag{2.1}$$

where l is the length of the armature in metres, r is the radius in metres and ω is the angular velocity in radians/second. The induced e.m.f. thus varies with the angular position of the coil in the same way as the flux density, as shown in Fig. 2.6. The action of the commutator is to invert the negative half cycles of the coil e.m.f., so the terminal voltage is given by

$$e = 2\omega lr \,|B(\theta)|, \tag{2.2}$$

as shown in Fig. 2.6. This is the instantaneous voltage, and the average value is

$$e_{av} = \frac{1}{\pi} \int_0^\pi e \, d\theta = \frac{1}{\pi} \int_0^\pi 2\omega l r B \, d\theta. \qquad (2.3)$$

Now $lr \, d\theta = da$, an element of area of the armature surface. Equation 2.3 therefore becomes

$$e_{av} = \frac{2}{\pi} \omega \int_0^\pi Blr \, d\theta = \frac{2}{\pi} \omega \int_{\theta=0}^{\theta=\pi} B \, da$$

$$= \frac{2}{\pi} \Phi \omega \text{ volts} \qquad (2.4)$$

where Φ is the magnetic flux entering half the armature surface from one field pole, i.e. Φ is the pole flux. It may be noted that equation 2.4 is still obtained if the field is not purely radial or uniform along the length of the armature; the integration is simply more complicated.

If a current i_c flows in the coil, there is a force on each conductor given by equation 1.14:

$$F = Bli_c \text{ newtons.}$$

The torque on the armature is therefore

$$T = 2Fr = 2Bli_c r \text{ newton metres.} \qquad (2.5)$$

The action of the commutator is to reverse the direction of i_c every half revolution; since the direction of B also reverses, the torque is unidirectional, and may be written as

$$T = 2ilr \, |B(\theta)|, \qquad (2.6)$$

where i is the current in the armature circuit. The average torque is therefore

$$T_{av} = \frac{1}{\pi} \int_0^\pi 2ilrB \, d\theta$$

$$= \frac{2}{\pi} \Phi i \text{ newton metres.} \qquad (2.7)$$

ARMATURE WINDINGS AND COMMUTATOR

The generated voltage of the simple d.c. machine is unidirectional, but far from constant. The performance can be improved by adding more conductors and commutator segments. Consider the effect of a second coil at right angles to the first (Fig. 2.7). The coil induced e.m.f. waveforms are shown in Fig. 2.8, together with the e.m.f. e appearing at the terminals. There is an improvement in the output waveform, but each coil is now only used for half the time.

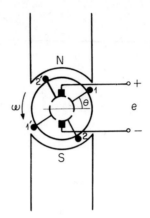

Fig. 2.7

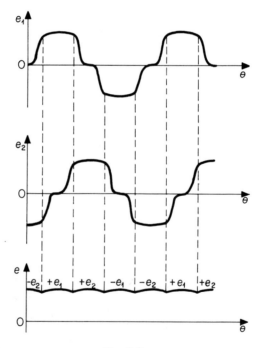

Fig. 2.8

In practical armature windings the coils and commutator segments are inter-connected so that the conductors carry current all the time, and there are usually several coils in series between the brushes. Fig. 2.9 shows how four coils and four commutator segments may be used to achieve this result; note that the current entering the armature divides into two parallel paths, and there are two coils in series in each path. A large number of coils and commutator segments are generally used, and each coil may have several turns to increase the generated voltage. There are many possible winding arrangements, which are described in standard texts such as Clayton and Hancock [1]; but the details do

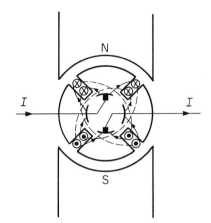

Fig. 2.9

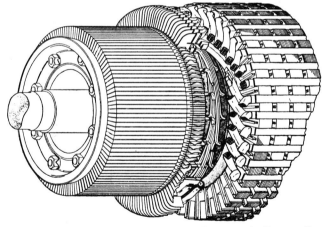

Fig. 2.10. Reproduced from *Electrical Machines* by A. Draper (Longmans, 2nd edition).

not concern us here. The construction of a typical d.c. machine armature is shown in Fig. 2.10, and it may be represented schematically by the diagram of Fig. 2.11. The function of any winding arrangement is to interconnect the coils and the commutator segments in such a way that all the conductors under one pole carry current in the same direction at all times, regardless of the motion of the armature. Notice that the current in any one armature conductor must reverse when the conductor passes the magnetic neutral axis between the poles. The currents in the armature coils are therefore alternating quantities, and the iron core of the armature is invariably laminated to reduce eddy current losses.

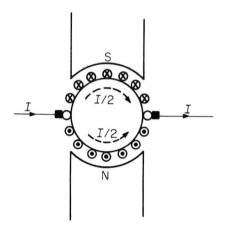

Fig. 2.11

GENERAL EQUATIONS OF THE D.C. MACHINE

In any armature winding there will be groups of coils in series between the brushes; the induced e.m.f.s will be additive, and if there are n conductors in series between the brushes the average induced e.m.f. will be

$$e_{av} = \frac{2n}{\pi} \Phi \omega \text{ volts.} \tag{2.8}$$

Since these n conductors each carry the same current, the average torque will be

$$T_{av} = \frac{2n}{\pi} \Phi i \text{ newton metres.} \tag{2.9}$$

More than two field poles may be employed, simply by repeating the N–S sequence as many times as desired round the periphery of the armature, with a corresponding modification of the armature winding; the fundamental principles remain unchanged. Thus a four-pole machine may be represented by the schematic diagram of Fig. 2.12. With a sufficiently large number of conductors, the generated voltage and the torque will be very nearly constant; we can

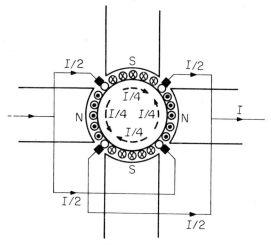

Fig. 2.12

generalize equations 2.8 and 2.9 to give

$$e_a = K_a \Phi \omega \text{ volts,} \qquad (2.10)$$

$$T = K_a \Phi i_a \text{ newton metres.} \qquad (2.11)$$

In these equations, e_a is the steady e.m.f. generated by the armature; i_a is the armature current, and the constant K_a is a property of the particular armature winding. In the derivation of these equations we have not assumed that ω or i_a is a constant quantity, and they hold for transient as well as steady-state conditions.

Equations 2.10 and 2.11 are the fundamental equations of the machine, and we shall shortly deduce some of the interesting and useful characteristics of d.c. machines from them. Their simplicity is striking, and it conceals the inherent complexity of the commutation process. The action of a well-designed armature winding and commutator is to convert the alternating quantities in the armature coils into perfectly steady quantities at the brush terminals. The coils necessarily possess inductance, and the reversal of current in an inductive circuit is accompanied by an induced e.m.f. or 'reactance voltage' which can cause sparking at the commutator. D.c. machines are therefore usually fitted with auxiliary poles (known as interpoles) to improve the commutation. These poles are placed mid-way between the main poles, and are wound with coils connected in series with the armature; their function is to induce an e.m.f. which opposes the reactance voltage in the armature coils undergoing commutation. The precise details of the commutation process are still imperfectly understood in spite of extensive research, and there remains an element of empiricism in the design of d.c. machines.

FIELD SYSTEM AND MAGNETIZATION CURVE

Fig. 2.13 is a schematic diagram of the structure of a d.c. machine without interpoles. The armature conductors operate in a magnetic field produced by coils wound on the field poles, and these coils are termed the field or excitation winding of the machine. A current i_f flowing in the field winding will produce a pole flux Φ, as shown. With no armature current flowing, Φ will be a function of

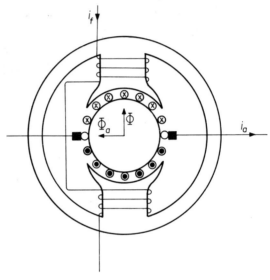

Fig. 2.13

i_f only, and we may put $\Phi = \Phi(i_f)$; a graph of Φ against i_f (or $N_f i_f$, the field ampere turns) is known as the magnetization curve of the machine. At constant speed, equation 2.10 gives $e_a \propto \Phi$; a graph of e_a against i_f for zero armature current is known as the open-circuit characteristic of the machine, and this has the same shape as the magnetization curve. Fig. 2.14 shows a typical curve, and it will be seen that the middle portion is practically linear. In this region the iron is unsaturated, and the air-gap reluctance is dominant in the magnetic circuit. The reluctance of the iron path increases rapidly with saturation, and this explains the shape of the curve for high values of field current. There is usually some remanent magnetization of the iron, which accounts for the departure from linearity at low values of field current. Fig. 2.14 is actually an over-simplification, for the magnetic circuit exhibits hysteresis; the curve for decreasing excitation will be slightly different from the curve for increasing excitation. The open-circuit characteristic may be regarded as the mean of the two curves.

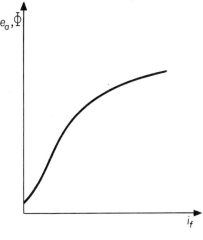

Fig. 2.14

LINEAR APPROXIMATION

It is usual to operate the machine below the knee of the magnetization curve, and for this region it is a reasonable approximation to put

$$\Phi = K_f i_f. \tag{2.12}$$

The fundamental equations for this linear model then become

$$e_a = K i_f \omega \text{ volts,} \tag{2.13}$$

$$T = K i_f i_a \text{ newton metres,} \tag{2.14}$$

where $K = K_a K_f$. These are the equations normally used in the analysis of systems containing d.c. machines.

ARMATURE REACTION

A current i_a flowing in the armature will produce a flux Φ_a, at right angles to Φ; this is known as the armature reaction flux, and by itself it will produce no torque or e.m.f. If the magnetic circuit were linear, there would be no interaction between Φ and Φ_a; in practice, Φ_a may cause local saturation of the magnetic circuit, and this will reduce the value of Φ for a given field current i_f. Thus $\Phi = \Phi(i_f, i_a)$, and in some applications this non-linear dependence of Φ on i_f and i_a must be considered. It is often sufficient, however, to ignore the effect of armature reaction and to take equations 2.13 and 2.14 as the basic machine equations. The armature reaction flux can adversely affect the commutation, especially when the armature current changes rapidly. To overcome this difficulty, d.c. machines are sometimes fitted with compensating windings. These take the form of conductors embedded in slots in the field pole faces; they are connected in series with the armature, but carry current in the opposite direction so as to cancel the armature reaction flux.

D.C. MACHINE ACTION IN TERMS OF MAGNETIC FORCES

The concept of armature reaction suggests an alternative physical picture for the mechanism of torque production in the d.c. machine. The existence of an armature reaction flux implies magnetization of the armature iron, which may be represented by N and S poles, and the resultant magnetic field produced by the armature and field poles is shown in Fig. 2.15. From the Maxwell stress

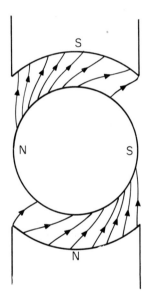

Fig. 2.15

concept (or the properties of magnetic poles) it is clear that there will be a torque on the armature tending to rotate its poles into alignment with the field poles. The armature winding and commutator, however, ensure that the magnetic axis of the armature remains fixed in space while the armature material revolves; a steady torque is therefore developed, which is unaffected by the rotation of the armature.

SLOTTED ARMATURE

In practice, as will be seen from Fig. 2.10, the armature conductors are placed in slots in the armature core. This profoundly alters the electromagnetic action of the machine, for the magnetic field in a slot is relatively weak (see Fig. 1.33) and the force on a conductor is reduced in consequence. It may be shown (Carter [2], Hague [3]) that the reduction in the conductor force is exactly compensated by forces acting on the slotted iron structure of the armature; the total torque is still given by equation 2.11. A qualitative explanation is that the magnetic field pattern around the armature (Fig. 2.15) is, on average, the same whether or not the conductors are placed in slots, and the torque produced by the Maxwell stresses will therefore be unchanged.

The e.m.f. generated by the machine is still given by equation 1.22, as may be seen by applying Faraday's law to an armature coil; the change of flux through the coil during one revolution of the armature is the same regardless of whether the coil sides are in slots or on the surface. The flux cutting rule $e = Blu$ will give different results if the field in the vicinity of the conductor is used for B; but the dangers inherent in the use of this rule have already been mentioned (see section 1.3).

An important principle is thus established: the basic machine equations are the same for a slotted armature as for a smooth cylindrical armature, provided that the total pole flux Φ (and hence the average value of B in the air gap) is unchanged. This result will be used in later chapters to simplify the analysis of a.c. machines.

2.2 Energy conversion and losses

Generators convert energy from mechanical to electrical form; motors perform the inverse operation, and the efficiency of energy conversion is often an important consideration. Since the efficiency is directly related to the energy loss in the machine, the sources of loss are matters of considerable importance to the machine designer. The machine user also needs to be aware of these losses, and it is useful to discuss them briefly before examining the other characteristics of d.c. machines.

GENERATORS AND MOTORS: SIGN CONVENTIONS

A d.c. machine with its armature connected to a steady voltage source V_a is shown symbolically in Fig. 2.16. The armature circuit (conductors, commutator and brushes) will have a resistance represented by R_a, and the field winding will have a resistance R_f. If the steady generated voltage E_a is less than V_a, the source will supply power to the machine, which therefore acts as a motor. If we choose the direction of current flow so that the armature current I_a is positive under these conditions, then

$$V_a = E_a + R_a I_a. \tag{2.15}$$

If $E_a > V_a$, the source will absorb power from the machine, which therefore acts as a generator. With the opposite direction for positive I_a, we now have

$$E_a = V_a + R_a I_a. \tag{2.16}$$

Generator and motor action differ only in the directions of current and torque, and we adopt the convention that both the torque and the armature current will

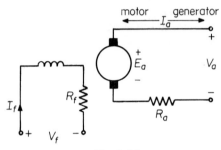

Fig. 2.16

be positive when the machine is operating as a motor. The correct armature voltage equation is therefore equation 2.15, *not* equation 2.16; negative values of I_a and T will indicate that the machine is operating as a generator.

LOSSES AND EFFICIENCY

The losses in a d.c. machine are essentially the same whether the machine operates as a generator or a motor, and motoring operation will be assumed for the rest of this section. Consider a d.c. machine with its armature connected to a voltage source V_a as shown in Fig. 2.16. With steady-state conditions the basic machine equations are

$$T = K_a \Phi I_a, \tag{2.10}$$

$$E_a = K_a \Phi \omega, \tag{2.11}$$

and we also have the motor armature equation

$$V_a = E_a + R_a I_a. \tag{2.15}$$

Multiplication of equation 2.10 by ω, and equations 2.11 and 2.15 by I_a gives

$$\omega T = K_a \Phi I_a \omega, \tag{2.17}$$

$$V_a I_a = E_a I_a + R_a I_a^2$$
$$= K_a \Phi \omega I_a + R_a I_a^2 . \tag{2.18}$$

From equations 2.17 and 2.18,

$$V_a I_a = \omega T + R_a I_a^2 , \tag{2.19}$$

showing that the electrical input power to the armature is divided between the gross mechanical power ωT and the resistive loss $R_a I_a^2$. The mechanical output from the motor shaft will be less than ωT, because some of the torque T (known as the gross, or electromagnetic, torque) will be absorbed in rotational losses. The flow of energy through the machine may be traced as follows:

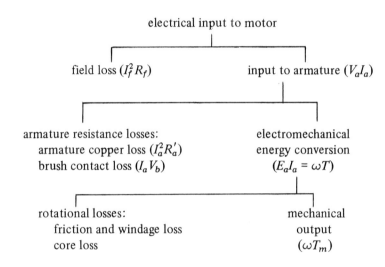

Components of loss

The losses in the machine comprise the field and armature resistance losses and the rotational losses. Although the armature resistance loss is usually represented by the expression $R_a I_a^2$, it should strictly be separated into the two components shown in the diagram because the brush contact resistance is non-linear. The voltage drop between the brushes and the commutator segments has an approximately constant value V_b over a wide range of currents, with a typical value of 2 volts for a pair of normal carbon brushes.

The two components of rotational loss are quite different in character. 'Friction and windage' accounts for all the mechanical and aerodynamic losses associated with the rotation of the armature, and varies roughly as the square of the speed. 'Core loss' includes the eddy-current and hysteresis loss in the laminated iron armature core, together with a similar surface loss in the field poles. The latter is a consequence of placing the armature conductors in slots; as the slots move past the field poles, the resulting local variations in the magnetic flux density produce eddy-current and hysteresis losses in the pole faces. A laminated construction is often used to minimize the eddy-current component of pole-face loss. The total core loss varies in a complex way with the speed, armature current and field flux.

Efficiency

An important property of an electrical machine is its efficiency under specified operating conditions. The efficiency η is defined in the usual way as

$$\eta = \frac{\text{output power}}{\text{input power}} \qquad (2.20)$$

$$= \frac{\text{input power} - \text{losses}}{\text{input power}}$$

$$= 1 - \frac{\text{losses}}{\text{input power}}. \qquad (2.21)$$

Because of the difficulty of measuring input and output power accurately, the direct evaluation of efficiency implied by equation 2.20 is seldom used; instead the losses are determined (usually from a number of different tests) and the efficiency is calculated from equation 2.21.

2.3 D.c. generators

As sources of d.c. power, d.c. generators have been largely replaced by controlled semiconductor rectifiers and a detailed treatment of their characteristics would be out of place in this book. Two types of generator, however, are worthy of mention: the separately excited generator, in which the field winding is supplied from a separate power source; and the shunt generator or dynamo, which supplies its own excitation power. Only steady-state operation is considered in this section; transient conditions are treated in section 2.5.

SEPARATELY EXCITED GENERATOR

When the machine is used as a source of power it is driven at a constant speed ω, and the field winding is connected to a voltage source V_f (Fig. 2.16). There will be a constant flux Φ, and the armature generated voltage has a constant value E_a given by equation 2.11. With no load connected to the armature terminals, $V_a = E_a$, and

$$V_a = K_a \omega \Phi. \tag{2.22}$$

The relationship between the open-circuit terminal voltage V_a and the field current I_f is given by the open-circuit characteristic (Fig. 2.14). In the linear region of the characteristic we may put

$$V_a = E_a \approx K \omega I_f = \frac{K\omega}{R_f} V_f, \tag{2.23}$$

showing that the output voltage is approximately proportional to the input voltage or current. The excitation power $V_f I_f$ is only a few per cent of the output power $V_a I_a$ when the machine is connected to a load; a d.c. generator may be regarded as a power amplifier, and has been widely used as such in control schemes such as the Ward–Leonard system described in section 2.4.

A separately excited machine may also be used to generate a voltage proportional to speed. Equation 2.10 shows that the generated voltage e_a is proportional to the speed ω if the field flux Φ is held constant. This may be done either by passing a constant current through the field winding, or by using permanent magnets to produce the field flux. Permanent magnet machines of this kind are known as tachogenerators, and they are widely used as speed-sensing elements in automatic control systems.

SHUNT GENERATOR

Instead of being connected to a separate voltage source, the field of a shunt generator is connected in parallel with the armature, so that the armature itself supplies the excitation current (Fig. 2.17). Frequently there is a variable resistance in series with the field, and the total resistance will be denoted by R_f.

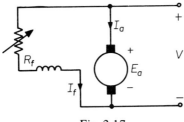

Fig. 2.17

The operation of a shunt generator may be understood by drawing on the open-circuit characteristic of the machine a straight line representing the voltage/current characteristics of the resistance R_f (Fig. 2.18). The point at which this line intersects the open-circuit curve gives the no-load terminal voltage of the machine, for the current flowing in the resistance is then just

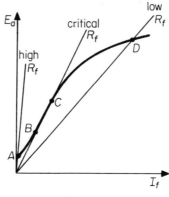

Fig. 2.18

equal to the field current required to maintain that value of terminal voltage. When the resistance is high, the line intersects the curve at a point such as A, and the armature voltage is very small. As the value of R_f is decreased, the point of intersection moves up the curve until a critical value of R_f is reached; for lower values of R_f the operating point is at a position such as D in the saturation region. When R_f is equal to the critical value the generator is unstable, and very small changes in conditions can move the operating point from B to C. Shunt generators are normally designed to work well into the saturation region, so that the operating point is stable and the output voltage substantially constant.

2.4 D.c. motors

The great virtue of the d.c. motor is that three useful operating characteristics may be obtained by using the field winding in different ways. In the separately excited motor the armature and field windings are supplied from independent voltage sources, and the speed of the motor may be controlled by varying the voltage applied to the armature. The shunt motor has its field connected in parallel with the armature, which gives an almost constant speed over a wide range of loads. The series motor has its field connected in series with the armature, and this gives a characteristic in which the speed is inversely related to the torque load on the motor.

The essential features of d.c. motor performance may be deduced fairly readily if the following idealizing assumptions are made:

(a)   the armature resistance R_a is neglected;
(b)   rotational losses are neglected;
(c)   magnetic non-linearity is ignored;
(d)   the motor operates under steady-state conditions.

As will be seen in the following discussion, these assumptions are not always appropriate; they greatly simplify the analysis, but their use must always be recognized and the implications appreciated.

IDEAL MOTOR CHARACTERISTICS

With the motor connected to a voltage source V_a (Fig. 2.16) the armature circuit equation is

$$V_a = E_a + I_a R_a.$$
[2.15]

The power loss in the armature resistance R_a is generally small in comparison with the input power to the motor, at least for machines with power ratings above 1 kW. It follows that the voltage drop $I_a R_a$ is usually small in comparison with V_a, and equation 2.15 may be written as

$$V_a \approx E_a.$$
(2.24)

In the steady state, therefore, the operating conditions of the machine are determined by the fact that the generated e.m.f. (or 'back e.m.f.', as it is usually termed in a motor) must be approximately equal to the armature supply voltage. Since the back e.m.f. E_a is related to the speed and the field flux by the equation

$$E_a = K_a \Phi \omega,$$
[2.11]

then

$$V_a \approx K_a \Phi \omega,$$

or $$\omega \approx \frac{V_a}{K_a \Phi}.$$
(2.25)

This equation is the basis of motor speed control, for it shows that the speed varies directly with the supply voltage V_a and inversely with the flux Φ. The other quantity of interest is the torque given by

$$T = K_a \Phi I_a.$$
[2.10]

If we idealize the machine by putting $\Phi = (K/K_a)I_f$, these equations become

$$\omega \approx \frac{V_a}{K I_f},$$
(2.26)

$$T = K I_f I_a.$$
(2.27)

Various methods of connecting the machine impose constraints between V_a, I_f and I_a, and we now explore some of the characteristics that may be obtained in this way.

SEPARATELY EXCITED MOTOR

With separate excitation (Fig. 2.16), I_f and V_a are controlled independently; the speed is given explicitly by equation 2.26 and the torque by equation 2.27. A motor is usually designed for a certain nominal speed at nominal values of armature voltage and field current; we examine the effect on the speed of varying the field current I_f and the armature voltage V_a from these nominal values.

Consider first the effect of varying the field current I_f. The normal value of field current will usually take the iron near to magnetic saturation. If the current I_f is increased above its normal value, saturation occurs; there is no longer a linear relationship between Φ and I_f, and equation 2.26 is not valid. Equation 2.25 shows that the speed varies inversely with the flux Φ, and the flux cannot change appreciably once the iron is saturated. Thus speed control is only possible if the field current is reduced below its normal value – a process known as field weakening. This causes the motor speed to rise above its nominal value, as shown by equation 2.26. The speed range obtainable by field weakening is rather restricted, for I_f cannot be reduced without limit. Equation 2.27 shows that the armature current I_a must increase inversely with I_f to maintain the torque, and commutation difficulties arise when the armature current is large and the field flux small.

Since field control can only give a limited rise of speed above the nominal value, armature voltage control must be used for speeds below the nominal value. Equations 2.25 and 2.26 show that the speed is nearly proportional to the armature voltage (for a constant field current), and a very wide speed range may be obtained by varying V_a. This linear relationship between speed and voltage is an important characteristic of the d.c. machine, which gives it a pre-eminent position in speed control systems (both manual and automatic), with sizes ranging from a few watts to tens of megawatts. Field current and armature voltage control are often combined in applications which require exceptionally large speed ranges.

WARD–LEONARD SPEED CONTROL SYSTEM

In this system (Fig. 2.19) a separately excited motor is supplied with a constant field current I_m. The armature supply is obtained from a separately excited generator driven at a constant speed ω_g by another motor (usually an a.c. induction motor). We have seen that the armature voltage of a separately excited generator is approximately proportional to its field current:

$$E_g = K_g \omega_g I_g. \tag{2.28}$$

Since the motor speed is proportional to the applied armature voltage, it is also approximately proportional to the generator field current:

$$\omega_g \approx \frac{E_g}{K_m I_m} = \frac{K_g \omega_g}{K_m I_m} I_g. \tag{2.29}$$

The Ward–Leonard system has been widely used for motor speed control, because the power required to control the field of the generator is only about

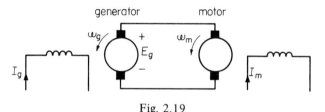

Fig. 2.19

1% of its output. The generator is thus acting as a d.c. power amplifier, and the cascade connection of several generators can give very high power gains. Unfortunately the dynamic performance of cascaded generators is rather poor, as will be shown in section 2.5, and the modern trend is to replace the generator by a controlled semiconductor rectifier operating from an a.c. supply. There are, however, still some applications which require d.c. generators. When the d.c. motor is subjected to large suddenly-applied loads (e.g. the main drive motors of a steel rolling mill), a large pulse of current is taken from the d.c. source supplying the motor armature. If this source is a controlled rectifier, a correspondingly large pulse of current will be taken from the a.c. mains, which may cause an unacceptable drop in the supply voltage. The severity of the pulse load on the mains can be reduced by using a motor-generator set in place of the rectifier. During the period of the pulse the speed of the set is allowed to fall, so that some of the energy required by the main motor is extracted from the stored rotational energy of the motor-generator set. A flywheel may be added to enhance this effect, and the arrangement is then known as a Ward–Leonard–Ilgner system.

SHUNT MOTOR

In this arrangement (Fig. 2.17) the field current I_f is obtained from the armature supply voltage V through a resistance R_f (internal winding resistance plus external variable resistance). Thus $I_f = V/R_f$, and equations 2.26 and 2.27 become

$$T = \frac{K}{R_f} VI_a,$$ (2.30)

$$\omega \approx \frac{R_f}{K}.$$ (2.31)

Thus the speed is independent of V, and varies directly with R_f. Torque and speed are still independent, and the shunt motor is essentially a constant speed machine. As with the separately excited machine, the range of speed variation obtainable by field weakening (increasing R_f) is limited, and the variable field resistance is normally used only for small variations from the nominal speed. If the effects of armature resistance are included, equation 2.31 becomes

$$\omega = \frac{R_f}{K} \left\{ 1 - \frac{R_a R_f T}{K V^2} \right\},$$ (2.32)

and the torque–speed curve is shown in Fig. 2.20. Some measure of speed control is possible by inserting additional resistance in series with the armature.

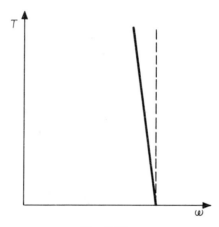

Fig. 2.20

Variation of this resistance gives a family of torque–speed curves, as shown in Fig. 2.21. If the torque–speed characteristic of the load is plotted on the same graph, its intersection with the motor torque–speed curve gives the speed at which the motor drives the load. The loaded speed of the motor therefore falls with increasing R_a. This is not a good method of speed control, because of the power loss in R_a and the dependence of the speed on the load torque. It does, however, permit the speed of a shunt motor to be reduced below the nominal

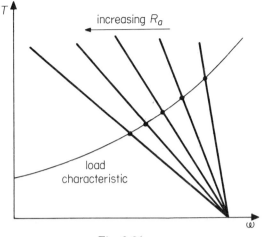

Fig. 2.21

value; field control, as we have seen, is only useful for raising the speed above this value.

SERIES MOTOR

If the field winding consists of a few turns capable of carrying the full armature current, it may be connected in series with the armature (Fig. 2.22). This introduces the constraint $I_f = I_a$ and if resistance is neglected we have, from equation 2.25 and 2.27,

$$I = \sqrt{(T/K)}, \tag{2.33}$$

$$\omega = \frac{V}{\sqrt{(KT)}}. \tag{2.34}$$

Torque and speed are no longer independent, and the torque–speed characteristic is shown in Fig. 2.23. This is a useful characteristic for a traction motor; the machine develops high accelerating torque at low speeds, and as the speed rises the torque falls until it is just sufficient to maintain the speed of the vehicle. Series d.c. motors are used in battery electric vehicles and electric trains. If the magnetic circuit is laminated to reduce eddy currents, a series d.c. motor will operate quite well from a single-phase a.c. supply. Universal motors of this kind are widely used in portable power tools and domestic appliances such as food mixers. The reason for using the relatively expensive construction of a commutator machine is that much higher speeds are possible than with induction motors, giving a larger power output from a given size. Also, the torque–speed .

characteristic of the series motor is more suitable than the induction motor characteristic for these applications. Speed control of the series motor may be achieved by variation of the supply voltage V; equation 2.34 shows that, for a given torque, the speed ω is directly proportional to V.

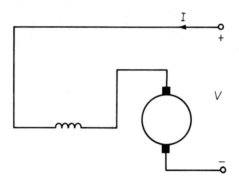

Fig. 2.22

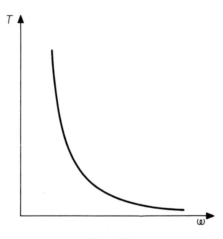

Fig. 2.23

An important characteristic of the series motor is the high speed attained when the machine is lightly loaded; the current in the series field is low, and a high rotational speed is needed to generate the required back e.m.f. in the weak field flux. With a small machine the windage and friction torque is sufficient to limit the no-load speed to a safe value, but a large series motor must never be started without a load or the speed will rise to a very high value and the armature may burst under the rotational stresses. For this reason an auxiliary shunt winding is sometimes added to limit the no-load speed.

COMPOUND MACHINES

A d.c. motor is sometimes built with both shunt and series field windings (Fig. 2.24), giving a characteristic in between that of a shunt motor and a series motor, as shown in Fig. 2.25.

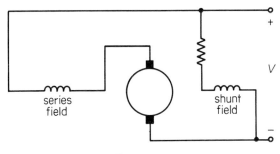

Fig. 2.24

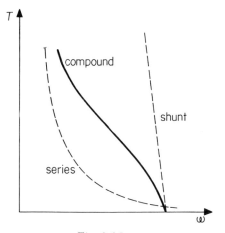

Fig. 2.25

STARTING OF D.C. MOTORS

We have seen that the steady-state operation of d.c. motors is governed by the condition that the armature back e.m.f. E_a is approximately equal to the supply voltage V_a. This condition does not hold when the motor is started from rest, for E_a is initially zero; if the armature were connected directly to the supply, a very large current would flow, limited only by the small armature resistance R_a. This current may damage the commutator, and a variable resistance is normally connected in series with the armature during starting; its value is progressively reduced as the armature runs up to speed.

The field current of a shunt or separately excited motor should always be set to its maximum value for starting, to give maximum starting torque with the available armature current and to give a rapid build-up of the back e.m.f. as the armature accelerates. It is dangerous to attempt to start a motor with the field winding disconnected, because the residual flux may give enough torque to accelerate the armature, which will attain a dangerously high speed in order to generate the required back e.m.f. For the same reason, the field supply to a motor must never be disconnected while the machine is running; the armature supply must always be switched off first. These precautions do not, of course, apply to series motors, but the danger of starting a series motor without a mechanical load has already been mentioned.

2.5 Dynamic characteristics of d.c. machines

DYNAMIC EQUATIONS

When a d.c. machine is not operating under steady-state conditions, it is necessary to take account of the inductance of the armature and field circuits and the inertia of the armature. With the circuit of Fig. 2.26, and the idealized machine equations, we have:

$$T = Ki_f i_a \tag{2.13}$$

$$e_a = Ki_f \omega \tag{2.14}$$

$$v_f = R_f i_f + L_f \frac{di_f}{dt} \tag{2.35}$$

$$v_a = e_a + R_a i_a + L_a \frac{di_a}{dt} \tag{2.36}$$

$$T_m = T - J \frac{d\omega}{dt}, \tag{2.37}$$

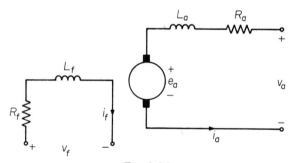

Fig. 2.26

where T_m is the mechanical output torque and J is the armature moment of inertia.

All quantities may now be time-varying, and the electrical quantities are written with lower-case letters to emphasize this. Note that there is no mutual inductance coupling between the field and armature circuits because the magnetic axes are at right angles; also that the machine is a unilateral device – the field current affects the armature circuit, but the converse is not true. In the mechanical equation of motion, rotational losses have been ignored.

Because of the product terms appearing in equations 2.13 and 2.14, specific problems often yield non-linear differential equations which may not be soluble analytically; analogue or digital computation must then be used. In many applications, however, one of the quantities in the product is a constant, and the solution of the problem is quite straightforward. Two simple examples will illustrate the methods.

TRANSIENT PERFORMANCE OF D.C. GENERATORS

Consider a d.c. generator driven at a constant speed ω, with the armature on open circuit. Suppose that a steady voltage V_f is suddenly applied at time $t = 0$ to the field winding; the field current is given by equation 2.35, and the solution is

$$i_f = \frac{V_f}{R_f} (1 - e^{-t/\tau_f}), \tag{2.38}$$

where $\tau_f = L_f/R_f$ is the field time constant. The generated e.m.f., given by equation 2.14, is then

$$e_a = K\omega i_f = \frac{K\omega V_f}{R_f} (1 - e^{-t/\tau_f}). \tag{2.39}$$

After a sufficient length of time this will attain the steady value

$$E_a = \frac{K\omega V_f}{R_f}. \tag{2.40}$$

Suppose now that a steady voltage V_a is suddenly applied to the armature terminals; the armature current is given by equation 2.36, with the solution

$$i_a = \frac{V_a - E_a}{R_a} (1 - e^{-t/\tau_a}), \tag{2.41}$$

where $\tau_a = L_a/R_a$ is the armature time constant. Note that there is no reflected voltage in the field circuit due to current in the armature circuit.

Cascade connection of d.c. generators

Typical values of time constants for a d.c. machine are $0.5 - 1$ s for τ_f and $0.1 - 0.2$ s for τ_a. It is the large value of the field time constant τ_f that limits the dynamic performance of d.c. generators, particularly when machines are connected in cascade. Fig. 2.27 shows two generators running at constant speed,

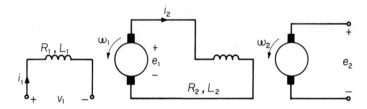

Fig. 2.27

with the armature of the first machine connected to the field of the second. The equations for the machines are

$$v_1 = R_1 i_1 + L_1 \frac{di_1}{dt},$$

$$e_1 = K_1 \omega_1 i_1,$$

$$v_2 = e_1 = R_2 i_2 + L_2 \frac{di_2}{dt},$$

$$e_2 = K_2 \omega_2 i_2.$$

Combining these gives

$$v_1 = \frac{R_1 R_2}{K_1 K_2 \omega_1 \omega_2} \left\{ e_2 + (\tau_1 + \tau_2) \frac{de_2}{dt} + \tau_1 \tau_2 \frac{d^2 e_2}{dt^2} \right\}, \qquad (2.42)$$

where $\tau_1 = L_1/R_1$ and $\tau_2 = L_2/R_2$.

If v_1 is a voltage step of magnitude V applied at time $t = 0$ and $\tau_1 = \tau_2 = \tau$, the solution of this second-order differential equation is

$$\frac{e_2}{E_2} = 1 - e^{-t/\tau} (1 + t/\tau), \qquad (2.43)$$

where E_2 is the final steady-state value of e_2 given by

$$E_2 = \frac{K_1 K_2 \omega_1 \omega_2}{R_1 R_2} V. \qquad (2.44)$$

Similarly, for n machines in cascade, all having the same field circuit time constant τ, the response to a step input is given by

$$\frac{e_n}{E_n} = 1 - e^{-t/\tau} \left\{ 1 + t/\tau + \frac{(t/\tau)^2}{2!} + \ldots + \frac{(t/\tau)^{n-1}}{(n-1)!} \right\}. \qquad (2.45)$$

The responses for $n = 1, 2, 3,$ and 4 are shown in Fig. 2.28, and it is clear that an increasingly sluggish response is the penalty for the high power gain of cascaded machines. This problem is overcome very ingeniously in the cross-field

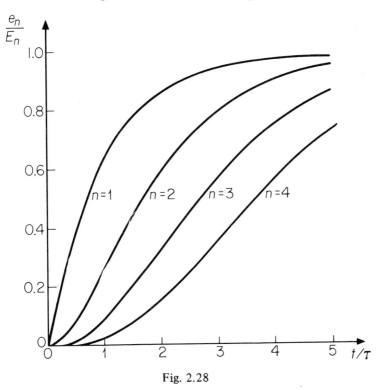

Fig. 2.28

machines or rotary amplifiers, which are effectively two cascaded machines combined together, with two sets of brushes on a single commutator; the armature reaction flux from one pair of brushes acts as the field flux for the second pair, thus eliminating one of the field time constants.

Starting of d.c. motors

As a second example of d.c. machine dynamics, consider the starting process of a shunt or separately excited motor. We assume a constant field current I_f, and no mechanical load so that $T_m = 0$. At time $t = 0$, a constant supply voltage V_a is suddenly connected to the armature. If the total armature circuit resistance is R_a, and the inductance may be neglected, then

$$V_a = e_a + R_a i_a$$
$$= K\omega I_f + R_a i_a. \tag{2.46}$$

Also
$$J\frac{d\omega}{dt} = T = K I_f i_a, \tag{2.47}$$

so that
$$V_a = K I_f \omega + \frac{R_a J}{K I_f}\frac{d\omega}{dt}. \tag{2.48}$$

The solution to this equation is

$$\omega = \Omega(1 - e^{-t/\tau_{em}}), \tag{2.49}$$

where $\Omega = V/K I_f$ is the final steady-state speed, and the time constant τ_{em} is given by

$$\tau_{em} = \frac{R_a J}{(K I_f)^2}. \tag{2.50}$$

This quantity is termed the electromechanical time constant, and it is generally much longer than the armature time constant τ_a. If the armature inductance L_a is not neglected, the starting process is described by the second-order differential equation

$$V_a = \frac{L_a J}{K I_f}\frac{d^2\omega}{dt^2} + \frac{R_a J}{K I_f}\frac{d\omega}{dt} + K I_f \omega. \tag{2.51}$$

Provided that $\tau_{em} \gg \tau_a$, the solution to this equation may be divided into two parts (Majmudar [4]):

(a) $\qquad t \ll \tau_{em}; \qquad \omega \approx 0, \quad i_a = \frac{V_a}{R_a}(1 - e^{-t/\tau_a}).$

(b) $\qquad t \gg \tau_a; \qquad \omega = \Omega(1 - e^{-t/\tau_{em}}), \quad i_a = \frac{V_a - K I_f \omega}{R_a}.$

Part (a) describes an electrical transient, where the armature current rises exponentially towards a limiting value V_a/R_a before the armature has time to accelerate. Part (b) describes a mechanical transient, where the armature speed rises exponentially towards its final limiting value Ω, and the armature current is only limited by the resistance R_a; the current is changing too slowly for the inductance L_a to have any significant effect. Thus the electrical and mechanical transients can be treated independently, provided that the two time constants are widely different. This is an instance of an important separation principle, which applies to many transient problems in electrical machines.

Problems

2.1 If the rotational losses of a d.c. shunt motor are constant, prove that the efficiency of the motor will be a maximum when the armature current is such that the armature resistance loss $I_a^2 R_a$ is equal to the sum of the rotational loss and the field resistance loss.

A 500 V d.c. shunt motor has a field winding resistance of 1 000 Ω and a rated full-load output power of 10 kW. If the rotational losses amount to 250 W and the efficiency is a maximum at full load, calculate this efficiency and the value of the armature resistance.

2.2 If the field and armature windings of a d.c. machine carry alternating currents with r.m.s. values I_f and I_a respectively, show that the average torque developed by the machine is given by

$$T = KI_f I_a \cos \phi$$

where ϕ is the phase angle between the currents and K is the machine constant.

A series motor is connected to an alternating voltage supply of r.m.s. value V. Show that the average torque is given by

$$T = \frac{KV^2}{(R + K\omega_r)^2 + X^2}$$

where K is the machine constant, ω_r is the armature angular velocity, R is the total series resistance and X is the total series reactance of the windings at the frequency of the supply. Obtain the corresponding torque expression for the same motor operating from a d.c. supply and discuss the difference between a.c. and d.c. operation.

2.3 The hoisting cable of a crane is wound onto a drum, and a d.c. shunt motor drives the drum through a reduction gearbox. The crane is used to lift a load at a steady speed, and the speed is controlled by varying a resistance R in series with the armature. The internal armature resistance may be neglected.

Show that the hoisting speed u is given by the expression

$$u = u_0 - A WR$$

where u_0 is the no-load speed, W is the weight of the load and A is a constant. Also show that the efficiency of the system is u/u_0. Rotational losses in the motor and the gearbox may be neglected, and the cable winds onto the drum at a constant radius.

2.4 A 200 V d.c. shunt motor has a no-load speed of 100 rad/s and its armature resistance is 1 Ω. Rotational losses are negligible. The motor drives a water pump, and the normal torque load on the motor is 40 N m. A fault in the water system causes the torque load on the motor to fall suddenly to 20 N m.

Calculate:

(a)   the normal armature current;
(b)   the normal speed at which the motor drives the pump;
(c)   the motor armature current just after the fault has occurred;
(d)   the final armature current;
(e)   the final motor speed.

If the rotating parts have a moment of inertia of 0.1 kg m^2 and the inductance of the armature may be neglected, obtain the differential equation which governs the speed of the motor after the fault.

2.5 A separately excited d.c. motor has a constant field current I_f. When a voltage v is applied to the armature a current i will flow; if the motor is unloaded, the resulting torque will accelerate the armature. Neglecting rotational losses, show that

$$v = \frac{(KI_f)^2}{J} \int i \, dt + R_a i + L_a \frac{di}{dt} \, ,$$

where J is the moment of inertia and K is the machine constant. Hence show that the machine is equivalent to a series combination of resistance R_a, inductance L_a and capacitance $C_m = J/(KI_f)^2$. If a constant voltage V is applied for a sufficiently long time, the current will be zero and the energy stored in the capacitance will be $\frac{1}{2}C_m V^2$. Prove that this is equal to the rotational energy of the armature.

References

[1] CLAYTON, A. E. and HANCOCK, N. N. (1959). *The performance and design of direct current machines,* 3rd edition. Pitman, London.
[2] CARTER, G. W. (1967). *The electromagnetic field in its engineering aspects,* 2nd edition. Longmans, London.
[3] HAGUE, B. (1962). *The principles of electromagnetism applied to electrical machines.* Dover, New York.
[4] MAJMUDAR, H. (1969). *Introduction to electrical machines.* Allyn and Bacon, Boston.

CHAPTER 3

Alternating Current Systems

With d.c. machines, the voltage generated by an individual armature coil is an alternating quantity which is rectified mechanically by the commutator. If we dispense with the commutator and revert to the slipring model of Fig. 2.1, we have a rudimentary a.c. generator. Practical a.c. generators are essentially simpler than their d.c. counterparts and are more easily designed in very large sizes; but a more important reason for their adoption is the possibility of using transformers to raise the voltage level for power transmission over long distances, and then to reduce it again for domestic or industrial consumption.

In nearly all applications, alternating current has advantages over direct current. The usefulness of the transformer is an obvious point, but it is the induction motor more than any other device which has vindicated the alternating current system. Induction motors are cheap, efficient and robust; they supply most of the motive power for industry, and they are used in many domestic appliances. The other main type of a.c. machine is the synchronous machine; most a.c. generators are of this kind, and in large sizes the synchronous motor is a strong rival to the induction motor. These machines are treated in chapters 4, 5 and 6, but it is first necessary to consider some of the properties of a.c. systems and transformers.

3.1 Generation of sinusoidal alternating voltages

The e.m.f. generated in a coil depends only on the rate of change of the flux linking the coil; it is of no consequence whether the coil moves in the field of fixed magnetic poles, or the poles move and the coil remains stationary. Usually, it is more convenient to have the active coils stationary and the field poles rotating, as shown in the simple model of Fig. 3.1. This is always done with large machines, to avoid the transfer of large amounts of power through sliprings. In the simple model, the rotor is magnetized by a field coil (or 'excitation winding'),

with the field current supplied via brushes and sliprings (not shown in Fig. 3.1).
The stationary part of the machine (the 'stator') carries a single turn armature
coil, and an e.m.f. will be induced in this coil when the rotor moves. If the

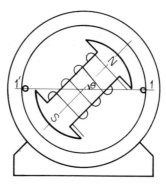

Fig. 3.1

rotor angular velocity is ω, the induced e.m.f., which has already been calculated
for the d.c. machine, is

$$e = 2Blr\omega, \tag{2.1}$$

where l is the length of a coil side and r is the radius. The flux density B is the
value at the right-hand coil side, which is displaced by an angle θ from the
magnetic axis of the rotor. Obviously B is a periodic function of θ (of period 2π),
and by suitably shaping the poles it can be arranged that

$$B = B_m \cos \theta. \tag{3.1}$$

If θ has some value ϕ_0 at time $t = 0$, and the angular velocity ω is a constant,
then $\theta = \omega t + \phi_0$ and equation 3.1 becomes

$$B = B_m \cos(\omega t + \phi_0). \tag{3.2}$$

The induced e.m.f. is now a sinusoidal* alternating quantity, given by

$$e = 2lr\omega B_m \cos(\omega t + \phi_0)$$
$$= E_m \cos(\omega t + \phi_0), \tag{3.3}$$

where $E_m = 2lr\omega B_m$. Sine waves are distinguished from all other periodic
functions by the fact that the steady-state response of any linear electric circuit
to a sine wave of voltage is also sinusoidal. It is therefore advantageous to
generate alternating voltages in the form of sine waves for general transmission
and distribution, and generators are normally designed to do this.

* Strictly speaking, this is a cosinusoidal quantity; but the term 'sinusoidal' will
    be used to describe a sine or a cosine function.

3.2 Polyphase systems

Suppose that a second armature coil is added to the simple generator, at right angles to the first, as shown in Fig. 3.2. If the flux density at coil side α is $B_m \cos \theta$, the corresponding value at coil side β is $B_m \cos(\theta - \pi/2)$. The generated voltages are then given by

$$
\left.
\begin{aligned}
e_\alpha &= E_m \cos(\omega t + \phi_0), \\
e_\beta &= E_m \cos(\omega t + \phi_0 - \pi/2).
\end{aligned}
\right\}
\tag{3.4}
$$

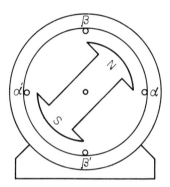

Fig. 3.2

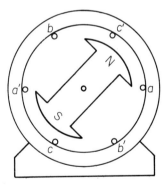

Fig. 3.3

These voltages have the same frequency but different phase angles; the two armature coils are known as phases, and the voltages are the phase voltages. This two-phase system is a rather special case, and the general symmetrical m-phase (or 'polyphase') system is obtained from m coils arranged symmetrically round the armature. Thus for a three-phase generator (Fig. 3.3) we have

$$\left.\begin{aligned}
e_a &= E_m \cos(\omega t + \phi_0), \\
e_b &= E_m \cos(\omega t + \phi_0 - 2\pi/3, \\
e_c &= E_m \cos(\omega t + \phi_0 - 4\pi/3).
\end{aligned}\right\} \tag{3.5}$$

These voltages may be represented by the phasor diagram of Fig. 3.4, and the corresponding waveforms are shown in Fig. 3.5. The generator is said to be balanced when all the phase voltages have the same amplitude, as they do here.

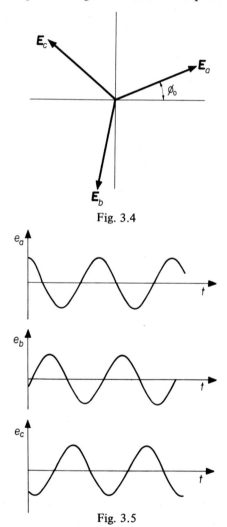

Fig. 3.4

Fig. 3.5

THREE-PHASE SYSTEMS

Most a.c. machines require balanced polyphase currents for satisfactory operation. (A notable exception is the small single-phase induction motor, discussed in section 6.5.) In principle any number of phases upwards of two could be used, with machines of appropriate design; in practice three phases are almost universally used for economic reasons. The economics of a.c. power transmission systems may be compared on the basis of the same maximum voltage between conductors and the same total I^2R-loss in all the conductors; a three-phase system using three conductors requires less total conductor material than any other number of phases, including single phase (Waddicor [1]).

The three armature coils of a three-phase generator may be represented in a circuit diagram by three voltage generators or sources. There are two ways in which these sources may be connected together so as to transmit power with less than six conductors. In the star or wye (Y) connection of Fig. 3.6a, there is a

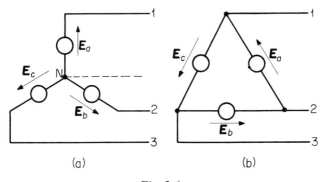

(a)   (b)

Fig. 3.6

choice of a three-wire or four-wire system, depending on whether the star (or neutral) point N is connected to a fourth line. The mesh or delta (Δ) connection of Fig. 3.6b is possible in a balanced system because $e_a + e_b + e_c = 0$ at all instants of time (this follows at once from the phasor diagram; the sum of the three phasors is zero). Only a three-wire system is possible with a delta-connected source. In each case the individual sources are known as phases. If we consider only three-wire systems, a three-phase load may be connected to the lines in two ways (Fig. 3.7). The impedances Z_a, Z_b and Z_c form the phases of the load; if they are equal the load is balanced.

When the load is star connected (Fig. 3.7a), the current in each phase of the load is equal to the current in the corresponding line. The voltage across a phase is not, however, equal to the voltage between a pair of lines. When the source

and load are balanced, the voltage phasor diagram takes the symmetrical form
shown in Fig. 3.8a, from which we obtain the relationship

$$V_{\text{line}} = \sqrt{3}\, V_{\text{phase}}. \tag{3.6}$$

With a delta-connected load (Fig. 3.7b), the voltage across each phase is equal to
the voltage between the corresponding pair of lines. It is now the phase and line

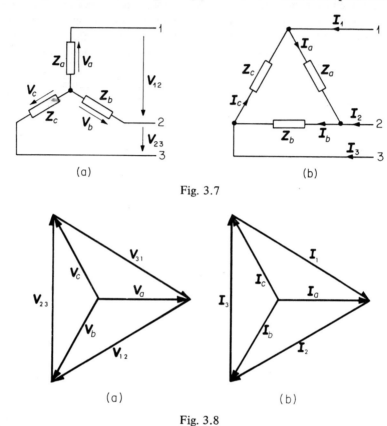

Fig. 3.7

Fig. 3.8

currents which are unequal, and the current phasor diagram for a balanced
system takes the form of Fig. 3.8b; this gives the relationship

$$I_{\text{line}} = \sqrt{3}\, I_{\text{phase}}. \tag{3.7}$$

It will be seen that star and delta connection are duals, with voltage in one
analogous to current in the other. Unbalanced systems may be handled by the
usual techniques of circuit analysis, with voltage or current sources representing
the phases of the generator. The star-delta $(T - \pi)$ transformation of circuit

theory is often useful for converting star loads to equivalent delta form, and vice versa. When the load is balanced, the relationship between the phase impedances for equivalent star and delta configurations is

$$Z_{\text{delta}} = 3 Z_{\text{star}}. \tag{3.8}$$

PHASE SEQUENCE

In the three-phase system so far considered, the voltages pass through their maximum positive values in the sequence $a \to b \to c$. This is the normal arrangement, and it is termed the positive phase sequence. If any pair of phases is interchanged the sequence will be reversed; thus interchanging e_b and e_c would give the system

$$e_{a'} = e_a = E_m \cos(\omega t + \phi_0),$$

$$e_{b'} = e_c = E_m \cos(\omega t + \phi_0 - 4\pi/3) = E_m \cos(\omega t + \phi_0 + 2\pi/3),$$

$$e_{c'} = e_b = E_m \cos(\omega t + \phi_0 - 2\pi/3) = E_m \cos(\omega t + \phi_0 + 4\pi/3).$$

The phase sequence for this system is $c' \to b' \to a'$, which is termed the negative sequence. It will be shown in chapter 4 that the direction of rotation of an a.c. motor depends on the phase sequence of the supply, and a positive sequence is normally assumed.

3.3 Transformers

A transformer is a pair of coils coupled magnetically (Fig. 3.9), so that some of the magnetic flux produced by current in the first coil links the turns of the second, and vice versa. The coupling can be improved by winding the coils on a

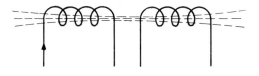

Fig. 3.9

common magnetic core (Fig. 3.10), and the coils are then known as the 'windings' of the transformer. We begin by considering an ideal transformer in which the coupling is perfect, i.e. the same core flux Φ links each turn of each winding.

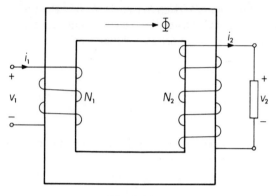

Fig. 3.10

THE IDEAL TRANSFORMER

Suppose that one winding (known as the primary) is connected to a voltage source v_1; it will draw a current i_1, and

$$v_1 = R_1 i_1 + N_1 \frac{d\Phi}{dt}. \tag{3.9}$$

If the other winding (known as the secondary) is connected to a load, a current i_2 will flow, and the terminal voltage v_2 is given by

$$v_2 = N_2 \frac{d\Phi}{dt} - R_2 i_2, \tag{3.10}$$

the negative sign arising from the different direction of current flow. If the windings have negligible resistance, then equations 3.9 and 3.10 reduce to

$$v_1 = N_1 \frac{d\Phi}{dt}, \tag{3.11}$$

$$v_2 = N_2 \frac{d\Phi}{dt}, \tag{3.12}$$

and these are the fundamental voltage equations of the ideal transformer. By division,

$$\frac{v_2}{v_1} = \frac{N_2}{N_1} = n, \tag{3.13}$$

where n is the turns ratio.

To find a relationship between i_1 and i_2, consider the magnetic circuit of the core. If S is the reluctance, then

$$S\Phi = F = N_1 i_1 - N_2 i_2. \tag{3.14}$$

The reluctance of the core is given by

$$S = \frac{l}{\mu_0 \mu_r A},$$ [1.75]

and if the relative permeability μ_r is high, S will be small. In an ideal transformer we postulate infinite permeability and therefore zero reluctance; equation 3.14 becomes

$$N_1 i_1 = N_2 i_2,$$ (3.15)

i.e. the primary ampere turns must balance the secondary ampere turns. The required current relationship is therefore

$$\frac{i_2}{i_1} = \frac{N_1}{N_2} = \frac{1}{n},$$ (3.16)

showing that the current transformation is the inverse of the voltage transformation. Equations 3.13 and 3.16 may be re-written in the form

$$v_2 = n v_1,$$ (3.17)

$$i_2 = \frac{1}{n} i_1.$$ (3.18)

On multiplying these equations together, we have

$$v_2 i_2 = v_1 i_1,$$ (3.19)

showing that the instantaneous power output is equal to the instantaneous power input.

With steady a.c. conditions we have the phasor equations

$$V_2 = n V_1,$$ (3.20)

$$I_2 = \frac{1}{n} I_1.$$ (3.21)

If an impedance Z_2 is connected to the secondary (Fig. 3.11), then $V_2/I_2 = Z_2$.

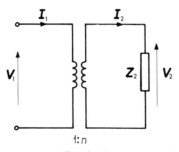

Fig. 3.11

Division of equation 3.20 by equation 3.21 gives

$$Z_2 = \frac{V_2}{I_2} = n^2 \frac{V_1}{I_1} .$$

Now V_1/I_1 is the impedance Z_1 presented by the primary terminals; hence

$$Z_1 = \frac{1}{n^2} Z_2 , \qquad (3.22)$$

and the transformer has the property of changing impedances. It is this property which makes the transformer so useful in electronic and communication circuits; in power circuits it is the voltage or current transforming property which is of interest.

THE REAL TRANSFORMER

In a real (as opposed to an ideal) transformer, the winding resistances are not zero; the magnetic coupling between the coils is not perfect; and the reluctance of the core is not zero. We now show that it is possible to represent the real transformer by an equivalent circuit consisting of an ideal transformer together with other elements which represent the imperfections. Since the transformer is a pair of coupled coils, we may use the coupled circuit equations derived in section 1.3, with the notation of Fig. 3.12:

$$\left. \begin{aligned} v_1 &= R_1 i_1 + L_1 \frac{di_1}{dt} - M \frac{di_2}{dt}, \\[2mm] v_2 &= -R_2 i_2 - L_2 \frac{di_2}{dt} + M \frac{di_1}{dt} . \end{aligned} \right\} \qquad (3.23)$$

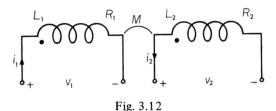

Fig. 3.12

The negative signs in these equations arise from the reversed direction of the secondary current i_2. The winding resistances occur explicitly in these equations but it is not obvious how the magnetic imperfections are to be included. It is necessary to re-arrange the inductive terms in equations 3.23 by introducing the concept of leakage inductance.

Leakage inductance

The reluctance of the magnetic circuit is finite, and the magnetic circuit law $NI = S\Phi$ implies that a current is required to set up the working flux in the transformer core. Thus if a current i_1 flows in the primary winding, with no current in the secondary, the flux linkage with the primary winding is given by

$$\psi_1 = L_1 i_1, \tag{3.24}$$

and the average flux per turn is

$$\Phi_1 = \frac{L_1 i_1}{N_1}. \tag{3.25}$$

Some of the flux produced by the primary current will also link the secondary winding; the flux linkage is given by

$$\psi_{21} = M i_1, \tag{3.26}$$

and the average flux per turn for the secondary is

$$\Phi_{21} = \frac{M i_1}{N_2}. \tag{3.27}$$

If the coupling were perfect the same flux would link each turn of each winding. On account of magnetic leakage (Fig. 3.13) the flux Φ_{21} linking the turns of the

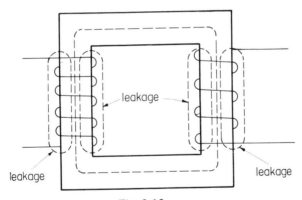

Fig. 3.13

secondary will be less than the flux Φ_1 linking the primary, and we may define a leakage flux Φ_{l_1} as the difference between these quantities:

$$\Phi_{l_1} = \Phi_1 - \Phi_{21}. \tag{3.28}$$

Since Φ_1 and Φ_{21} are both proportional to the current i_1 we may put

$$\Phi_{l_1} = \frac{l_1 i_1}{N_1}. \tag{3.29}$$

The quantity l_1 is termed the leakage inductance of the primary winding; its physical interpretation is that the flux produced by the current i_1 flowing in the inductance l_1 represents that portion of the primary flux which fails to link with the secondary.

The leakage inductance l_1 may be expressed in terms of L_1 and M, for we have

$$\Phi_{l_1} = \Phi_1 - \Phi_{21}$$
$$= \frac{L_1 i_1}{N_1} - \frac{M i_1}{N_2},$$

[3.28]

and equation 3.29 gives the result

$$l_1 = \frac{N_1 \Phi_{l_1}}{i_1} = L_1 - \frac{N_1}{N_2} M.$$

(3.30)

Similarly there is a secondary leakage flux given by

$$\Phi_{l2} = \Phi_2 - \Phi_{12} = \frac{L_2 i_2}{N_2} - \frac{M i_2}{N_1},$$

(3.31)

and a secondary leakage inductance

$$l_2 = L_2 - \frac{N_2}{N_1} M.$$

(3.32)

Equivalent circuit

The leakage inductances may be incorporated into the coupled circuit equations; from equations 3.30 and 3.32 we have

$$\left. \begin{aligned} L_1 &= l_1 + \frac{N_1}{N_2} M, \\ L_2 &= l_2 + \frac{N_2}{N_1} M, \end{aligned} \right\}$$

(3.33)

and equations 3.23 become

$$\left. \begin{aligned} v_1 &= R_1 i_1 + l_1 \frac{di_1}{dt} + \frac{1}{n} M \frac{d}{dt}(i_1 - n i_2), \\ v_2 &= -R_2 i_2 - l_2 \frac{di_2}{dt} + M \frac{d}{dt}(i_1 - n i_2), \end{aligned} \right\}$$

(3.34)

where $n = N_2/N_1$ as before. Equations 3.34 are the equations of the circuit shown in Fig. 3.14, which is the required equivalent circuit. This is a time–domain

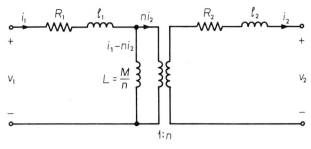

Fig. 3.14

circuit, valid for arbitrarily time-varying voltages and currents, subject only to the assumptions made in deriving the equations. These are:

(a)   capacitance effects, between turns and between the two windings, are negligible;

(b)   magnetic non-linearity is ignored;

(c)   iron losses are ignored.

We now consider steady-state sinusoidal operation and the representation of iron losses.

A.C. EQUIVALENT CIRCUIT OF THE TRANSFORMER

For sinusoidal a.c. operation we may draw an a.c. (frequency domain) circuit corresponding to Fig. 3.14, with inductances replaced by reactances and instantaneous quantities replaced by phasor (complex) quantities. The a.c. circuit is shown in Fig. 3.15, in which an additional resistance R_l has been connected in

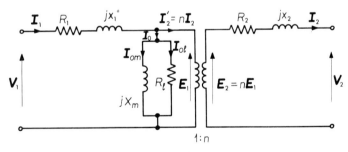

Fig. 3.15

parallel with the reactance jX_m. This resistance represents the iron loss in the core, and the reason for placing it in parallel with jX_m will be explained presently. Capacitance has again been neglected, so the circuit is only valid up to the low audio frequencies.

The equivalent circuit consists of an ideal transformer of ration $1 : n = N_1 : N_2$, together with elements which represent the imperfections of the real transformer. The voltage E_1 across the primary of the ideal transformer represents the voltage induced in the primary winding by the mutual flux Φ. This is the portion of the core flux which links both primary and secondary coils; the leakage flux associated with one winding does not link the other, and it is represented in the circuit by the leakage reactances jx_1 and jx_2. Similarly, the voltage E_2 across the secondary of the ideal transformer represents the voltage induced in the secondary winding by the mutual flux Φ. From equation 3.11 for the ideal transformer, we have

$$E_1 = j\omega N_1 \Phi, \tag{3.35}$$

showing that the induced voltage is in quadrature with the mutual flux. The magnitudes are related by

$$\Phi = \frac{E_1}{\omega N_1}, \tag{3.36}$$

showing that E_1 must vary in proportion to ω if the magnitude of the flux is to remain constant; this is the basis of the 'constant volts per cycle' rule for variable-frequency operation of iron-cored apparatus.

In the ideal transformer, the reluctance of the core is zero and there is an exact ampere-turn balance between the primary and secondary. In the real transformer, on the other hand, the reluctance of the core is finite, and the primary ampere turns $N_1 I_1$ must exceed the secondary ampere turns $N_2 I_2$ by an amount necessary to set up the mutual flux Φ in the core. When the secondary current I_2 is zero, the primary current has a finite value I_0, known as the no-load current. This current has a component I_{0l}, in phase with E, to supply the eddy-current and hysteresis losses in the core; and a component I_{0m}, known as the magnetizing current, to set up the core flux. The magnetizing current is, of course, the current flowing in the inductive reactance of the primary winding. From Fig. 3.15 and equation 3.35 we have

$$I_{0m} = \frac{E_1}{jX_m} = \frac{\omega N_1}{X_m} \Phi, \tag{3.37}$$

showing that the magnetizing current I_{0m} is in phase with the mutual flux Φ.

Core losses

It only remains to justify the use of the resistance R_l to represent iron losses. The hysteresis and eddy current components of iron loss depend on the frequency of the field and on the maximum value of the magnetic flux density in the core. For a given frequency, the mutual core flux Φ is proportional to the induced e.m.f. E_1, and the core loss may therefore be represented by the power loss in a resistance R_l connected across E_1. It may be shown (Jones [2]) that the value of R_l is only independent of the frequency and voltage if hysteresis losses are absent; since this is not the case, the correct value of R_l must be chosen for the particular operating conditions. With power transformers the frequency is fixed and E_1 does not vary by more than a few per cent when the primary supply voltage V_1 is held constant, so the variation in the value of R_l may be ignored.

Phasor diagram

When the secondary current I_2 is zero (i.e. the secondary is on open circuit) the primary current I_1 is just equal to I_0, the total current flowing in the shunt elements R_l and jX_m. When a load is connected to the secondary, a current I_2 flows, and the primary current is increased by an amount $I_2' = nI_2$. This is termed the load component of the primary current, or the secondary current *referred* to the primary, and the total primary current is given by

$$I_1 = I_0 + nI_2. \tag{3.38}$$

From the equivalent circuit, we also have

$$V_2 = E_2 - I_2(R_2 + jx_2)$$
$$= nV_1 - nI_1(R_1 + jx_1) - I_2(R_2 + jx_2). \tag{3.39}$$

These and other relationships are conveniently illustrated by the phasor diagram of Fig. 3.16. Equations 3.38 and 3.30 reduce to the ideal transformer equations when the no-load current I_0 is zero, and the leakage impedances $R_1 + jx_1$ and $R_2 + jx_2$ are zero. Instrument transformers are used to extend the

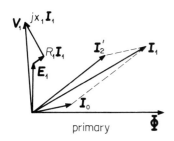
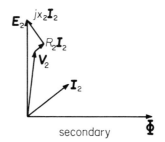

Fig. 3.16

voltage or current ranges of a.c. measuring instruments, and they are required to approximate as closely as possible to the ideal transformer; this is achieved by careful design, to minimize the unwanted terms in equations 3.38 and 3.39.

Simplification of the equivalent circuit

Practical iron-cored transformers are usually designed so that under normal working conditions the volt drop in R_1 and x_1 is small in comparison with V_1, and I_0 is small in comparison with the load current I_1. The shunt components R_l and X_m may then be transferred to the input terminals with very little loss of accuracy (Fig. 3.17a). The secondary quantities R_2 and x_2 may be replaced by

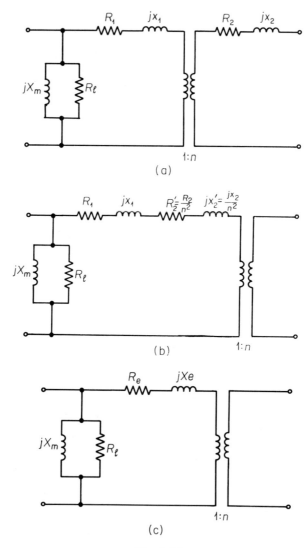

(a)

(b)

(c)

Fig. 3.17

equivalent quantities on the primary side, using the impedance transforming property of the ideal transformer, as shown in Fig. 3.17b. The actual primary quantities R_1 and x_1 can then be combined with the referred secondary quantities R'_2 and x'_2 to give the equivalent primary quantities R_e and X_e (Fig. 3.17c). This is the normal form of the equivalent circuit for small power transformers; it simplifies the analysis because the current I_0 is determined solely by the applied primary voltage V_1, and the parameters are readily determined from simple tests. Fig. 3.17 shows the elements referred to the primary side, and a 'mirror image' circuit is readily derived with all the elements referred to the secondary. This form is useful when the transformer is operated in an inverted mode, with a voltage applied to the secondary and a load connected to the primary.

DETERMINATION OF THE EQUIVALENT CIRCUIT PARAMETERS

The parameters of the approximate equivalent circuit are readily obtained from open-circuit and short-circuit tests.

In the open-circuit test one winding is left on open circuit and the normal voltage is applied to the other winding; only the small no-load current will be drawn from the supply. If the secondary is open-circuited the referred secondary current I_2 will be zero, and the equivalent circuit reduces to the form shown in Fig. 3.18. Measurements are made of the voltage V_o, the current I_o, and the

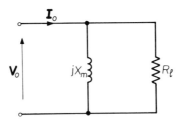

Fig. 3.18

input power P_o. The shunt elements X_m and R_l may then be determined from the relations

$$R_l = V_o^2/P_o, \tag{3.40}$$

$$Y_o = I_o/V_o, \tag{3.41}$$

$$X_m = 1/\sqrt{(Y_o^2 - 1/R_l^2)}. \tag{3.42}$$

For the short-circuit test, one winding is short circuited and the normal full-load current is allowed to flow in the other winding by connecting it to an adjustable low-voltage source. Thus if the secondary is short circuited, a short circuit will be reflected into the primary side of the equivalent circuit. The leakage impedance is thus placed in parallel with the magnetizing impedance; since these generally differ by at least two orders of magnitude the magnetizing impedance may be ignored, and the equivalent circuit takes the form shown in

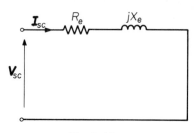

Fig. 3.19

Fig. 3.19. Measurements are made of the voltage V_{sc}, the current I_{sc} and the input power P_{sc}; the series elements are then determined from the relations

$$R_e = P_{sc}/I_{sc}^2, \tag{3.43}$$

$$Z_e = V_{sc}/I_{sc}, \tag{3.44}$$

$$X_e = \sqrt{(Z_e^2 - R_e^2)}. \tag{3.45}$$

In practice the open-circuit measurements are usually made on the low-voltage side of the transformer and the short-circuit measurements on the high-voltage side (where the current will be lower). This is done merely for convenience, to avoid using higher voltages and currents than necessary; the same equations apply, but for one test the element values will be referred to the primary and for the other they will be referred to the secondary. Conversion of secondary values to equivalent primary values (or vice versa) involves the turns ratio of the transformer; this is taken to be equal to the primary/secondary voltage ratio measured in the open-circuit test.

REGULATION AND EFFICIENCY

Because of the volt drop in the series impedances of the windings, the transformer secondary terminal voltage will vary with the load current. The *voltage regulation* is defined as

$$\epsilon = \frac{\text{(no-load voltage)} - \text{(full-load voltage)}}{\text{(full-load voltage)}} \times 100\%, \tag{3.46}$$

assuming a constant applied primary voltage.

The *efficiency* of the transformer is defined as

$$\eta = \frac{\text{output power}}{\text{input power}} \times 100\% \qquad (3.47)$$

$$= \left\{1 - \frac{\text{losses}}{\text{output} + \text{losses}}\right\} \times 100\%. \qquad (3.48)$$

The form of equation 3.48 for the efficiency is the one generally used, and the efficiency is determined from a measurement of losses. This is because the efficiency of a power transformer can be 99% or more; with this figure, an error of 10% in the loss measurement would give the same error in the efficiency as an error of 0.1% in the measurement of input and output power.

Problems

3.1 In a balanced three-phase four-wire system, show that no current will flow in the neutral conductor. Hence show that any balanced three-phase system (three-wire or four-wire) may be resolved into three separate single-phase systems with the voltages, currents and element values in the phases equal to the equivalent star values for the original system.

3.2 A three-phase load consists of three equal impedances of magnitude Z and phase angle ϕ. If the load is star connected, show that the *instantaneous* flow of power into the load is constant, with a value given by

$$P = \sqrt{3}\, VI \cos \phi$$

where V is the r.m.s. line voltage, I is the r.m.s. line current and $\cos \phi$ is the load power factor. Show that this expression also holds when the load is delta connected, and find the ratio of the line currents in the two cases.

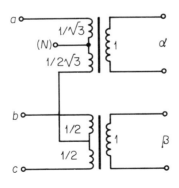

Fig. 3.20

3.3 The Scott connection of two transformers, shown in Fig. 3.20, is used for obtaining a two-phase supply α, β from a three-phase supply a, b, c with an optional neutral connection N. The transformers (which may be assumed to be ideal) have tapped primary windings, and the turns ratios indicated in the diagram. Construct a phasor voltage diagram for the system, and verify that the system will transform voltages from three to two phases and vice versa.

3.4 The ideal gyrator is a circuit element invented by Tellegen to complete the set of linear passive elements (resistor, capacitor, inductor, transformer, gyrator). The circuit symbol for a gyrator is shown in Fig. 3.21, and the defining equations are

$$v_1 = ki_2$$

$$v_2 = ki_1.$$

Compare the properties of the gyrator with those of an ideal transformer. If a capacitance C is connected to the output of the gyrator, what does the voltage-current relation at the input represent?

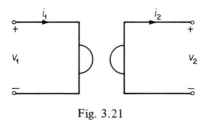

Fig. 3.21

3.5 A power transformer has the following nominal rating: primary 220 V, 50 Hz; secondary 660 V, 7.7 A, 5.1 kVA. Open and short circuit tests gave the following results. With the secondary winding an open circuit, measurements taken at the *primary* were 220 V, 1.18 A, 65.5 W; the secondary voltage was 669 V. With the primary winding short circuited, measurements taken at the *secondary* were 13.2 V, 7.70 A, 69.5 W.

   Calculate the parameters of the approximate equivalent circuit referred to the primary side of the transformer. Hence calculate (a) the secondary terminal voltage, (b) the efficiency, (c) the voltage regulation, when the transformer supplies its rated full-load secondary current of 7.7 A at unity power factor, with a primary supply voltage of 220 V.

3.6 In addition to the mains frequency of 50 Hz, a frequency of 400 Hz is commonly encountered in a.c. control systems. Consider two transformers of similar rating, one designed for operation at 50 Hz and the other for 400 Hz. When each transformer is operating at its rated voltage and frequency the peak core flux density is 1.3 T and the magnetizing current is 5% of the full-load primary current. The transformer cores are made from 4% silicon steel, with a lamination thickness appropriate to the frequency. By considering the core flux density, the magnetizing current and the iron losses, explain what will happen if (a) the 400 Hz transformer is operated at 50 Hz; (b) the 50 Hz transformer is operated at 400 Hz. In each case the normal rated voltage is applied to the primary of the transformer. Also explain why the 400 Hz transformer will be smaller and lighter than the 50 Hz transformer.

References

[1] WADDICOR, H. (1964). *Principles of electric power transmission*, 5th edition. Chapman & Hall, London.
[2] JONES, C. V. (1967). *The unified theory of electrical machines.* Butterworths, London.

CHAPTER 4

Introduction to A.C. Machines

The commutator in a d.c. machine performs a complex function which we have not attempted to analyse in detail; the performance equations, on the other hand, are relatively simple. With the a.c. synchronous and induction machines, the situation is reversed. The absence of a commutator simplifies both the structure and the detailed analysis; but the machine equations are more complex and the basic theory is conceptually more difficult. In this chapter we develop some of the principles which are common to all a.c. machines by considering the magnetic field set up by currents flowing in the windings. It is convenient to start with the a.c. generator introduced in the last chapter.

The simple model of a generator introduced in section 3.1 has a rotor with prominent poles (a 'salient pole' rotor). Many practical machines are in fact made with a rotor structure of this kind, but the general theory is quite complex and is best handled by the methods introduced in chapter 7. A simpler theory results if the rotor is cylindrical, giving a uniform air gap (Fig. 4.1), and this

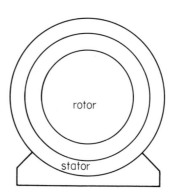

Fig. 4.1

form of construction is used for mechanical reasons in the high-speed turbine-driven generators of modern power stations. We have seen that a sinusoidal generated voltage is desirable, and this implies a sinusoidal variation of the flux density round the rotor when there is a single stator armature coil spanning a diameter. There are methods of arranging the coils in a practical armature winding to eliminate or reduce some of the harmonics when the flux density variation is not a pure sinusoid, but these are design details that do not concern us here. Whatever may be achieved by subtle design of the armature winding, it is generally necessary to make the flux density variation as nearly sinusoidal as possible. With a salient pole rotor this is accomplished by shaping the poles; with a cylindrical rotor it can only be done by distributing the conductors of the field winding in a particular way.

4.1 Distributed windings and the air-gap magnetic field

The coils forming the stator and rotor windings of a practical machine are usually arranged so that the conductors are distributed round the stator and rotor, instead of being concentrated at a number of points as they are in the simple models so far considered. To understand the operation of the machine it is necessary to calculate the magnetic field in the air gap from a knowledge of the current in the winding and the way in which the conductors are distributed; this may be done by an extension of the magnetic circuit concept.

M.M.F. OF A DISTRIBUTED WINDING

Fig. 4.2 shows a cross-section through a cylindrical-rotor machine, with conductors distributed round the stator and rotor surfaces. Application of Ampere's circuital law to the path $PQRS$ gives

$$\Sigma i = \oint H \, . \, ds = \int_P^Q H \, . \, ds + \int_Q^R H \, . \, ds + \int_R^S H \, . \, ds + \int_S^P H \, . \, ds. \qquad (4.1)$$

In this equation, Σi is the total current carried by the conductors within the boundary $PQRS$. If the permeability of the stator and rotor iron is high, the magnetic potential drops along the iron paths QR and SP will be negligible in comparison with the air-gap potential drops along PQ and RS. It is reasonable to assume that H is nearly uniform along a radial path in the air gap, so equation 4.1 becomes

$$\Sigma i = \int_P^Q H \, . \, ds + \int_R^S H \, . \, ds$$

$$= g \, H_r(\theta_2) - g \, H_r(\theta_1), \qquad (4.2)$$

where g is the radial length of the air gap, and H_r is the radial component of the magnetizing force. Usually, the magnetic field is nearly radial, so that $H_r \approx |H|$; we shall therefore drop the subscript r, with the understanding that the quantity H appearing in the equations is, strictly speaking, the radial component. At some

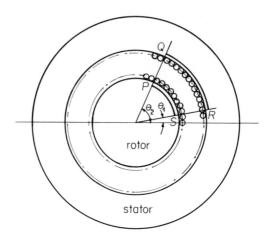

Fig. 4.2

point in the air gap, the value of H will be zero; let the corresponding value of θ be θ_0. At any other value of θ we have

$$\Sigma i = g\,H(\theta) - g\,H(\theta_0)$$

$$= g\,H(\theta), \tag{4.3}$$

where Σi is now the sum of the currents in the conductors between θ_0 and θ. This quantity Σi is known as the m.m.f. of a winding; since it depends on θ, it may be denoted by $F(\theta)$. Thus the radial component of the magnetic field depends on the air-gap length g and also on the quantity F, which specifies the way in which the winding conductors are distributed.

M.m.f. of a simple rotor winding

For a simple application of the winding m.m.f. concept, consider the rotor winding shown in Fig. 4.3. The conductors are distributed uniformly over two quadrants of the surface, with the current directed as shown. If each conductor carries a current i and there are n conductors per unit length of arc, the current contained between 0 and θ is

$$F(\theta) = \Sigma i = inr\theta \tag{4.4}$$

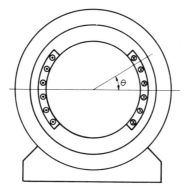

Fig. 4.3

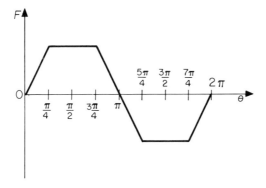

Fig. 4.4

where r is the radius of the rotor, and $|\theta| < \pi/4$. The complete m.m.f. distribution is shown in Fig. 4.4. The m.m.f. increases linearly until $\theta = \pi/4$, when a further increase in angle gives no further increase in the sum of the currents. This continues until $\theta = 3\pi/4$, when current of opposite sign is included; the m.m.f. then decreases linearly until $\theta = 5\pi/4$, and so on.

SINUSOIDALLY DISTRIBUTED WINDINGS

The requirement of a sinusoidal variation in the flux density round the rotor implies that the magnetizing force H in the air gap must also vary sinusoidally, and equation 4.3 shows that the m.m.f. $F(\theta)$ must be a sinusoidal function of θ. This will be the case if each conductor carries a current i, and the number of conductors between θ and $\theta + d\theta$ is given by the expression $k \cos \theta \, d\theta$. Then

$$F = \int_{\theta_0}^{\theta} ik \cos \theta \, d\theta$$

$$= ik \sin \theta , \qquad (4.5)$$

where $\theta_0 = 0$ to give $H(\theta_0) = 0$. A winding with a sinusoidal m.m.f. is known as a sinusoidally distributed winding; this is a theoretical ideal which can never be exactly realized in practice. The trapezoidal m.m.f. wave of Fig. 4.4 is a useful first approximation to a sine wave. Considerable improvement results from the stepped conductor distribution of Fig. 4.5, giving the waveform shown in Fig. 4.6. Note that the conductor distribution is a stepped approximation to the

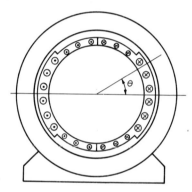

Fig. 4.5

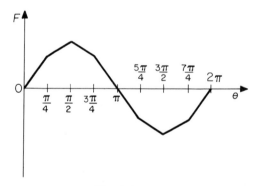

Fig. 4.6

ideal cosinusoidal distribution, and the integration of this wave gives the straight line segments of the resulting m.m.f. wave. An example of a practical machine winding with a stepped conductor distribution is given in section 4.7.

The concepts of a sinusoidally distributed winding and a sinusoidal magnetic field are of fundamental importance in a.c. machines. The requirement of a rotor winding of this kind arose from the need to generate sinusoidal voltages in the stator coils. If these coils form part of a distributed stator winding, the magnetic field produced by currents in this winding should have the same form as the rotor field; and this implies that the stator winding should also be sinusoidally distributed. Optimum machine performance is obtained with sinusoidal quantities, and by a happy coincidence this also gives the simplest mathematical treatment. An exactly sinusoidal distribution will therefore be assumed in developing the theory of a.c. machines.

4.2 Combination of sinusoidally distributed fields

We have seen that conductors distributed in a cosinusoidal manner will give rise to a sinusoidal m.m.f. wave. In developing the theory of a.c. machines it is more convenient to work with a cosinusoidal m.m.f. wave; this will result if the conductor density is $-k_1 \sin \theta$, for the m.m.f. is then

$$F_1 = \int_{\theta_0}^{\theta} -i_1 k_1 \sin \theta \, d\theta = i_1 k_1 \cos \theta, \tag{4.6}$$

where $\theta_0 = -\pi/2$ to satisfy the condition that $H(\theta_0) = 0$, and i_1 is the current flowing in each conductor. From equation 4.3 and the relation $B = \mu_0 H$ in the air gap we now have

$$B_1 = \mu_0 H_1 = \mu_0 \frac{F_1}{g} = \frac{\mu_0 k_1}{g} i_1 \cos \theta. \tag{4.7}$$

This may be written as

$$B_1 = B_{1m} \cos \theta, \tag{4.8}$$

where

$$B_{1m} = \frac{\mu_0 k_1}{g} i_1. \tag{4.9}$$

If the conductor distribution is shifted by an angle α, so that the m.m.f. is $F_1 = i_1 k_1 \cos(\theta - \alpha)$, the magnetic field is given by

$$B_1 = B_{1m} \cos(\theta - \alpha). \tag{4.10}$$

This sinusoidally distributed field pattern is illustrated in Fig. 4.7.

Suppose that there is a second sinusoidally distributed winding carrying a current i_2, which produces an m.m.f. $F_2 = i_2 k_2 \cos(\theta - \beta)$. This will give rise to a second magnetic field

$$B_2 = B_{2m} \cos(\theta - \beta), \tag{4.11}$$

where

$$B_{2m} = \frac{\mu_0 k_2}{g} i_2. \tag{4.12}$$

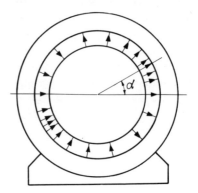

Fig. 4.7

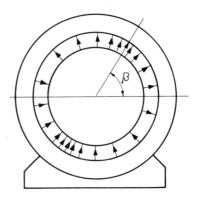

Fig. 4.8

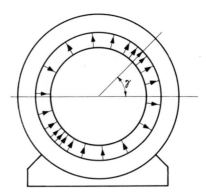

Fig. 4.9

This is illustrated in Fig. 4.8. The total magnetic field B is the sum of the separate fields:

$$B = B_1 + B_2$$

$$= B_{1m} \cos(\theta - \alpha) + B_{2m} \cos(\theta - \beta). \tag{4.13}$$

Since the sum of two sinusoids is another sinusoid, this may be written as

$$B = B_m \cos(\theta - \gamma), \tag{4.14}$$

and the magnetic field distribution is shown in Fig. 4.9.

VECTOR REPRESENTATION OF SINUSOIDAL FIELDS

It will be recalled that in a.c. circuit theory the manipulation of sinusoids is simplified by the use of rotating vectors or phasors. A similar device can be used with sinusoidally distributed magnetic fields. We represent a field of maximum value B_m by a radius vector of length B_m, pointing in the direction of the maximum field intensity. Thus in Fig. 4.10, the field B_1 is represented by a

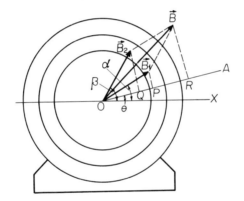

Fig. 4.10

vector $\vec{B}_1$ of length B_{1m}, making an angle α with the reference axis OX; and the field $\vec{B}_2$ is represented by a vector $\vec{B}_2$ of length B_{2m}, making an angle β with the reference axis. Let OA be a line making an angle θ with OX. The projection of $\vec{B}_1$ on OA is

$$OP = B_{1m} \cos(\alpha - \theta) = B_{1m} \cos(\theta - \alpha), \tag{4.15}$$

and the projection of $\vec{B}_2$ on OA is

$$OQ = B_{2m} \cos(\beta - \theta) = B_{2m} \cos(\theta - \beta). \tag{4.16}$$

Thus the value of the magnetic field at any angle θ in the air gap is equal to the projection of the respective vector onto a line making an angle θ with the reference axis. Now consider the vector sum $\vec{B}_1 + \vec{B}_2$. The projection of this vector onto OA is given by

$$OR = OP + PR$$

$$= OP + OQ$$

$$= B_{1m} \cos(\theta - \alpha) + B_{2m} \cos(\theta - \beta), \qquad (4.17)$$

and it therefore represents the total magnetic field B. The length of this vector is B_m, and it makes an angle γ with OX.

This procedure is exactly analogous to the phasor addition of sinusoidally time-varying voltages or currents. Following Chapman [1], we shall use the term 'spacer phasor' for the vector representing the sinusoidally distributed field, to avoid confusion with the magnetic field vector. Magnetic flux density is a vector quantity possessing magnitude and direction. In developing the theory of electrical machines we are mainly concerned with the radial component of the flux density vector, and in particular with the variation of the scalar magnitude of this component round the air gap. It is the spatial variation of a scalar magnitude which is described by the 'vector' introduced in this section.

4.3 Torque production from sinusoidally distributed windings

Suppose that the rotor and stator each have sinusoidally distributed windings. If the stator winding carries a current i_1, it will produce a magnetic field of the form

$$B_1 = B_{1m} \cos(\theta - \alpha), \qquad [4.10]$$

and the stator will behave like a permanent magnet with north and south poles. Likewise, if the rotor carries a current i_2 it will produce a magnetic field of the form

$$B_2 = B_{2m} \cos(\theta - \beta), \qquad [4.11]$$

and the rotor will also behave like a permanent magnet. The magnetic axes of the stator and rotor will be displaced by an angle $\delta_{12} = \alpha - \beta$, and there will be a torque on the rotor tending to pull its poles into alignment with the stator poles. This is illustrated in Fig. 4.11, where the bending of the lines of force is greatly exaggerated. The magnetic field cannot be purely radial, or there would be no torque on the rotor. This follows from the tangential Maxwell stress formula

$$t_s = \frac{B_n B_s}{\mu_0} ; \qquad [1.43]$$

if the field is purely radial, $B_s = 0$, and therefore $t_s = 0$. It is shown in Appendix A that a non-uniform radial magnetic field must be accompanied by a tangential component; when the stator and rotor axes are displaced, the radial and tangential components combine to give a field pattern of the form shown in Fig. 4.11.

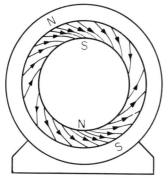

Fig. 4.11

The torque tending to align the magnetic axes of the stator and rotor may be calculated in a number of ways. Energy methods are commonly used (Chapman [1], Fitzgerald [2]), but a more direct method is to consider the force acting on the current in each element of the rotor surface*; this is merely an extension of the method already used for the d.c. machine. Let di be the current in an element of arc $d\theta$; this current will interact with the magnetic field B_1 of the stator to give a force

$$dF = l \times B_1 \, di, \tag{4.18}$$

where l is the axial length of the conductors in the arc $d\theta$. Only the radial component of B_1 will give rise to a tangential component of dF, and the contribution to the torque is thus

$$dT = -rlB_{1r} \, di, \tag{4.19}$$

where B_{1r} is the radial component of B_1, and r is the radius of the rotor. The negative sign arises from the convention that counter-clockwise torque is positive. Since the value of B_1 calculated from the winding m.m.f. is in fact the radial component (see section 4.1), equation 4.10 may be written in the form

$$B_{1r} = B_{1m} \cos(\theta - \alpha). \tag{4.20}$$

The current di is obtained from an equation corresponding to equation 4.6 for the rotor, giving

$$di = -i_2 k_2 \sin(\theta - \beta) \, d\theta. \tag{4.21}$$

---

* It is shown in Appendix A that this is equivalent to evaluating the Maxwell stress.

Since the rotor current i_2 is related to the maximum value B_{2m} of the rotor magnetic field by equation 4.12, the last equation may be written as

$$di = -\frac{g}{\mu_0} B_{2m} \sin(\theta - \beta)\, d\theta. \tag{4.22}$$

Substitution of these expressions for B_{1r} and di into equation 4.19 and integrating to obtain the total torque gives

$$T = \int_0^{2\pi} \frac{rlg}{\mu_0} B_{1m} \cos(\theta - \alpha) B_{2m} \sin(\theta - \beta)\, d\theta$$

$$= KB_{1m}B_{2m} \sin \delta_{12} \text{ newton metres,} \tag{4.23}$$

where $\qquad K = \dfrac{\pi rlg}{\mu_0} \quad \text{and} \quad \delta_{12} = \alpha - \beta. \tag{4.24}$

Equation 4.23 is an important result, which shows that the alignment torque T varies as the sine of the angular separation δ_{12} between the stator and rotor magnetic axes.

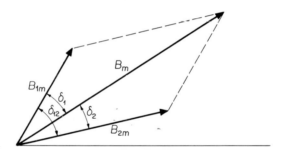

Fig. 4.12

The relation between the component fields B_1 and B_2 and the total field B is given by the space phasor diagram of Fig. 4.12. Application of the sine rule to this diagram gives

$$\frac{B_{1m}}{\sin \delta_2} = \frac{B_{2m}}{\sin \delta_1} = \frac{B_m}{\sin \delta_{12}} \tag{4.25}$$

The torque equation may therefore be written as

$$T = KB_{1m}B_{2m} \sin \delta_{12} = KB_{1m}B_m \sin \delta_1 = KB_{2m}B_m \sin \delta_2, \tag{4.26}$$

showing that the torque is proportional to the product of any two of the field components and the sine of the angle between them.

4.4 The rotating magnetic field

The equation

$$B = B_m \cos(\theta - \psi) \tag{4.27}$$

represents a sinusoidally distributed magnetic field with its axis inclined at an angle ψ. Suppose that this magnetic field is produced by a distributed winding on the stator, and that by some means the angle ψ is made to increase uniformly with time:

$$\psi = \psi_0 + \omega t, \tag{4.28}$$

where ω is a constant. For convenience, let $\psi_0 = 0$. The axis, and thus the sinusoidal magnetic field pattern, is rotating in the positive (counter-clockwise) direction with an angular velocity ω. The expression for the magnetic field becomes

$$\begin{aligned} B &= B_m \cos(\theta - \omega t) \\ &= B_m \cos(\omega t - \theta), \end{aligned} \tag{4.29}$$

and this is the equation of a rotating magnetic field. At any particular instant of time t, equation 4.29 shows that the field is sinusoidally distributed round the air gap, with its axis inclined at an angle ωt. The equation also shows that at any particular angle θ the field varies sinusoidally with time, but with a phase lag of θ. This is exactly analogous to a sinusoidal travelling wave of the form $\cos(x - ut)$, encountered in transmission line theory.

It has been seen that a magnetized rotor will experience a torque tending to align its axis with the magnetic axis of the stator. When the stator field rotates, the rotor will rotate in synchronism with it, and this is the basis of the synchronous machine which we study in chapter 5. An entirely different kind of machine results if the rotor does not carry fixed magnetic poles, but is made of conducting material or has conductors arranged in suitable closed circuits. If the rotor runs at a lower speed than the rotating field, the rotor will have currents induced in it by the relative motion of the field. These currents will interact with the field to produce a torque on the rotor, and this is the basis of the induction machine studied in chapter 6. We now consider how a rotating stator magnetic field may be produced from fixed windings.

PRODUCTION OF A ROTATING MAGNETIC FIELD

Suppose that the stator is provided with two sinusoidally distributed windings α and β, which are similar in all respects except that their axes are at $\theta = 0$ and $\theta = \pi/2$ respectively. This arrangement is termed a two-phase winding. If the windings carry currents i_α and i_β, the m.m.f. produced by each winding will be

$$\left.\begin{array}{l} F_\alpha = F_{\alpha m} \cos \theta, \\ F_\beta = F_{\beta m} \cos(\theta - \pi/2), \end{array}\right\} \tag{4.30}$$

where the amplitudes are given by

$$\left.\begin{array}{l} F_{\alpha m} = ki_\alpha, \\ F_{\beta m} = ki_\beta. \end{array}\right\} \tag{4.31}$$

The total m.m.f. is

$$F = F_\alpha + F_\beta = F_m \cos(\theta - \psi), \tag{4.32}$$

and the magnetic flux density is given by

$$B = \mu_0 H = \frac{\mu_0 F}{g}. \tag{4.33}$$

The amplitude F_m and space phase ψ of the resultant m.m.f. are given by the space phasor diagram of Fig. 4.13, from which

$$\left.\begin{array}{l} F_{\alpha m} = F_m \cos \psi, \\ F_{\beta m} = F_m \sin \psi. \end{array}\right\} \tag{4.34}$$

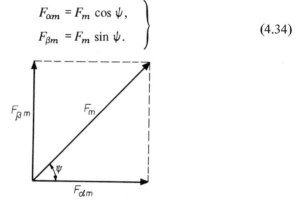

Fig. 4.13

For a pure rotating field we require F_m = constant and $\psi = \omega t$; from equations 4.34 and 4.31 we thus have

$$\left.\begin{array}{l} ki_\alpha = F_{\alpha m} = F_m \cos \omega t, \\ ki_\beta = F_{\beta m} = F_m \sin \omega t. \end{array}\right\} \tag{4.35}$$

Equations 4.35 will be true if the currents are given by

$$\left.\begin{array}{l} i_\alpha = I_m \cos \omega t, \\ i_\beta = I_m \sin \omega t = I_m \cos(\omega t - \pi/2). \end{array}\right\} \tag{4.36}$$

These are the equations of two-phase alternating currents, and it can be concluded that a rotating magnetic field will be produced by two-phase currents flowing in a two-phase winding. This result may also be obtained by direct substitution of equations 4.36 in equation 4.31. Then

$$B = \frac{\mu_0 F}{g} = \frac{\mu_0}{g}(kI_m \cos \omega t \cos \theta + kI_m \sin \omega t \sin \theta)$$

$$= B_m \cos(\omega t - \theta), \tag{4.37}$$

where
$$B = \frac{\mu_0 k}{g} I_m. \tag{4.38}$$

For simplicity, the two-phase winding has been chosen with the β phase leading the α phase by $90°$. It is customary to show the β phase lagging by $90°$, and this is the convention adopted in chapter 7. This makes no essential difference to the analysis; with the currents given by equation 4.36, the field would be $B = B_m \cos(\omega t + \theta)$, signifying rotation in the negative direction; reversing the sign of i_β (i.e. changing the phase from $90°$ lag to $90°$ lead) will restore the positive direction of rotation.

ROTATING MAGNETIC FIELD WITH A THREE-PHASE WINDING

A rotating magnetic field may likewise be produced from a symmetrical m-phase supply, provided that the armature also has an m-phase symmetrical winding. Thus for three phases, the m.m.f.s of the individual armature phases will be

$$\left. \begin{array}{l} F_a = i_a k \cos \theta, \\ F_b = i_b k \cos(\theta - 2\pi/3), \\ F_c = i_c k \cos(\theta - 4\pi/3). \end{array} \right\} \tag{4.39}$$

If the windings are supplied from a three-phase source, the currents will be

$$\left. \begin{array}{l} i_a = I_m \cos \omega t, \\ i_b = I_m \cos(\omega t - 2\pi/3), \\ i_c = I_m \cos(\omega t - 4\pi/3). \end{array} \right\} \tag{4.40}$$

The total m.m.f. is $F = F_a + F_b + F_c$, and the air-gap flux density is therefore

$$B = \frac{\mu_0 k I_m}{g} \{ \cos \omega t \cos \theta + \cos(\omega t - 2\pi/3) \cos(\theta - 2\pi/3) +$$

$$+ \cos(\omega t - 4\pi/3) \cos(\theta - 4\pi/3) \}$$

$$= \frac{\mu_0 k I_m}{g} \frac{1}{2} \{ \cos(\omega t - \theta) + \cos(\omega t + \theta) + \cos(\omega t - \theta) +$$

$$+ \cos(\omega t + \theta - 4\pi/3) + \cos(\omega t - \theta) + \cos(\omega t + \theta - 8\pi/3) \}$$

$$= B_m \cos(\omega t - \theta), \tag{4.41}$$

where
$$B_m = \frac{3\mu_0 k}{2g} I_m. \tag{4.42}$$

Reversal of the direction of rotation

Interchanging i_b and i_c would give

$$\left. \begin{array}{l} i_{a'} = i_a = I_m \cos \omega t, \\ i_{b'} = i_c = I_m \cos(\omega t - 4\pi/3) = I_m \cos(\omega t + 2\pi/3), \\ i_{c'} = i_b = I_m \cos(\omega t - 2\pi/3) = I_m \cos(\omega t + 4\pi/3). \end{array} \right\} \tag{4.43}$$

With these expressions for current, we obtain

$$B = B_m \cos(\omega t + \theta), \tag{4.44}$$

which represents a magnetic field rotating in the opposite direction. Thus the direction of rotation of the field may be reversed by reversing the phase sequence of the supply.

4.5 Voltage induced by a rotating magnetic field

Alternating currents flowing in suitable windings will produce a rotating magnetic field; we now calculate the voltage induced by the rotating field, and find a relationship between the voltage and current for each phase of the winding under balanced operating conditions.

A two-phase machine winding is the simplest, and this form of winding was used in the early days of a.c. systems. Since modern power systems use three phases for economic reasons, most industrial machines have three-phase windings. Domestic induction motors are usually single-phase machines, but these are special cases which are considered later in chapter 6. The two-phase machine is now confined almost entirely to a.c. servo systems (Taylor [3]); but the principles are virtually the same no matter how many phases are used to produce a uniform rotating field, and only the simpler two-phase model will be analysed in detail. The results may often be generalized to other numbers of phases by inspection, and it is shown in Appendix B that there is a mathematical transformation which enables the performance of a three-phase machine to be calculated from the analysis of an equivalent two-phase machine. This transformation holds for all conditions of operation: balanced or unbalanced, steady-state or transient; so there is no loss of generality in using a two-phase model.

VOLTAGE INDUCED IN A WINDING

Consider a single-turn coil in the air gap, as shown in Fig. 4.14, and a rotating magnetic field given by

$$B = B_m \cos(\omega t - \theta).$$ [4.37]

The flux linking this coil is given by

$$\Phi = \int_{-\pi/2}^{\pi/2} Blr \, d\theta \,,$$ (4.45)

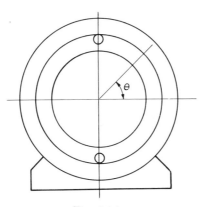

Fig. 4.14

where r is the radial distance of a coil side from the axis, and l is the length of each coil side. Evaluating the integral in equation 4.45 gives

$$\Phi = \int_{-\pi/2}^{\pi/2} B_m lr \cos(\omega t - \theta) \, d\theta$$

$$= 2rlB_m \sin(\omega t + \pi/2),$$ (4.46)

and the induced e.m.f. in the coil is thus

$$e = \frac{d\Phi}{dt} = 2rl\omega B_m \cos(\omega t + \pi/2).$$ (4.47)

This result could also be obtained from the flux cutting rule (equation 1.5); the velocity of the field relative to the conductor is $u = \omega r$, and the magnitude of the e.m.f. induced in each conductor is

$$e_c = Blu = rl\omega B_m \cos(\omega t + \pi/2). \tag{4.48}$$

In general, however, it is unwise to use the flux cutting rule in this way: see section 1.3.

If the single-turn coil is replaced by a sinusoidally distributed winding, the induced e.m.f. may be found by integration; for the α-phase winding the result is

$$e_\alpha = \pi r l k \omega B_m \cos(\omega t + \pi/2). \tag{4.49}$$

RELATIONSHIP BETWEEN PHASE VOLTAGE AND CURRENT

Equation 4.49 gives the e.m.f. induced in the α phase by the rotating magnetic field. When the field is itself produced by currents flowing in the two phases of the winding, there is an important relationship between the phase voltage and current. If the winding resistance and leakage reactance may be neglected, the terminal voltage v_α is equal to the induced e.m.f. e_α. The maximum flux density is related to the maximum current in the winding by the equation

$$B_m = \frac{\mu_0 k}{g} I_m, \tag{4.38}$$

and the terminal voltage is therefore given by

$$v_\alpha = \frac{\pi r l k^2 \mu_0}{g} \omega I_m \cos(\omega t + \pi/2). \tag{4.50}$$

The current flowing in this phase is

$$i_\alpha = I_m \cos \omega t. \tag{4.36}$$

Comparison of the last two equations shows that (a) the voltage leads the current by $\pi/2$ radians; (b) the amplitude of the voltage is proportional to the amplitude of the current; (c) the amplitude of the voltage is proportional to the angular frequency ω. In terms of phasors we have

$$V_\alpha = j\omega \left[\frac{\pi r l k^2 \mu_0}{g} \right] I_\alpha, \tag{4.51}$$

showing that this phase of the machine behaves as an inductance of value

$$L = \frac{\pi r l k^2 \mu_0}{g} \text{ henrys.} \tag{4.52}$$

It is readily shown that the terminal voltage of the β phase is given by

$$v_\beta = \frac{\pi r l k^2 \mu_0}{g} \omega I_m \sin(\omega t + \pi/2), \qquad (4.53)$$

while the current is

$$i_\beta = I_m \sin \omega t. \qquad (4.54)$$

Thus the β phase also behaves as an inductance of value L; the voltage and current have the same magnitudes as those in the α phase, but lag by an angle of $\pi/2$ radians. Similar results hold for a symmetrical m-phase machine; the currents and voltages have the same magnitudes in each phase, but there is a phase shift of $2\pi/m$ between adjacent phases.

RELATIONSHIP BETWEEN SPACE AND TIME PHASORS

Suppose that a machine has a winding on the stator which produces a rotating field of the form

$$B_1 = B_{1m} \cos(\theta - \omega t - \alpha), \qquad (4.55)$$

and another winding on the rotor which produces a rotating field of the form

$$B_2 = B_{2m} \cos(\theta - \omega t - \beta). \qquad (4.56)$$

These will combine to give a total field

$$B = B_m \cos(\theta - \omega t - \gamma), \qquad (4.57)$$

and at any instant of time the magnetic field components may be represented by the space phasors $\vec{B}_1$, $\vec{B}_2$ and $\vec{B}$. Fig. 4.15 shows the space phasor diagram for the instant $t = 0$, and the phasors may be imagined to rotate with angular velocity ω in the counter-clockwise direction. Each field component acting alone would

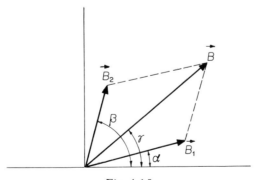

Fig. 4.15

induce e.m.f.s in the phases of the windings. Consider the e.m.f. e_1 induced in the α phase of the stator winding by the field component B_1. The previous analysis may be used if we let

$$\omega t' = \omega t + \alpha; \tag{4.58}$$

equation 4.55 then becomes

$$B_1 = B_{1m} \cos(\omega t' - \theta), \tag{4.59}$$

and equation 4.49 gives the induced e.m.f.

$$e_1 = KB_{1m} \cos(\omega t' + \pi/2) = KB_{1m} \cos(\omega t + \alpha + \pi/2) \tag{4.60}$$

where
$$K = \pi r l k \omega. \tag{4.61}$$

Similarly the e.m.f. e_2 induced by the field component B_2 will be

$$e_2 = KB_{2m} \cos(\omega t + \beta + \pi/2), \tag{4.62}$$

and the total e.m.f. e induced by the total field B will be

$$e = KB_m \cos(\omega t + \gamma + \pi/2). \tag{4.63}$$

These e.m.f.s may be represented by the time phasors E_1, E_2 and E shown in Fig. 4.16, which is drawn for the instant $t = 0$. Apart from a rotation of $90°$, this is similar to the space phasor diagram of Fig. 4.15; the lengths of the time phasors are proportional to the lengths of the corresponding space phasors, and the angles between the phasors are the same in the two diagrams.

   Currents must flow in the stator winding to produce the field B_1, and the phasor I_1 representing the current in the α phase will lag the e.m.f. E_1 by $90°$ as shown in Fig. 4.16. The relationship between the magnetic field components represented by Fig. 4.15 and the phase voltage and current represented by Fig. 4.16 is crucial to the theory of a.c. machines; it leads directly to the equivalent circuit of the synchronous machine in chapter 5, and in a more subtle way to the equivalent circuit of the induction machine in chapter 6.

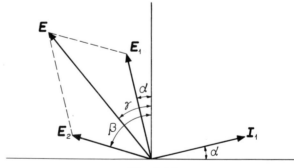

Fig. 4.16

4.6 Multi-pole fields

The machine windings so far considered produce two effective magnetic poles, and the field makes one revolution during one cycle of the a.c. supply. The length of arc between one pole and the next is known as a pole pitch, and we may therefore say that the field moves through two pole pitches during one cycle of the supply.

Suppose that we construct a machine having a four-pole rotor and a stator winding that produces a four-pole field (Fig. 4.17). During one cycle of the

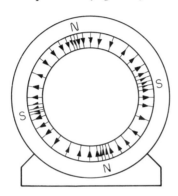

Fig. 4.17

supply the field will move through two pole pitches as before, but this only represents half a revolution. The speed of the rotating field is thus $\omega/2$ for a four-pole machine, and for a machine with $2p$ poles the speed will be ω/p. The magnetic field will still be sinusoidally distributed, and it will go through a complete cycle in two pole pitches regardless of the number of poles on the machine. It is convenient, therefore, to work in terms of an electrical angle θ_e which increases by 2π for each complete cycle of the field. If the actual angular position or 'mechanical angle' is θ, then the electrical angle is given by

$$\theta_e = p\theta. \tag{4.64}$$

This procedure may be justified as follows. The m.m.f. of a $2p$-pole two-phase winding will be given by

$$\left.\begin{array}{l} F_\alpha = ki_\alpha \cos p\theta, \\[2mm] F_\beta = ki_\beta \cos(p\theta - \pi/2). \end{array}\right\} \tag{4.65}$$

If a two-phase supply is connected to the winding, the currents will be

$$\left.\begin{array}{l} i_\alpha = I_m \cos \omega t, \\[2mm] i_\beta = I_m \cos(\omega t - \pi/2). \end{array}\right\} \tag{4.66}$$

The total m.m.f. is $F = F_\alpha + F_\beta$; from equations 4.65 and 4.66 this is

$$F = kI_m \cos(\omega t - p\theta)$$
$$= kI_m \cos p\left(\frac{\omega}{p}t - \theta\right) \tag{4.67}$$

When the supply angular frequency is ω, equation 4.67 shows that a $2p$-pole field will rotate with angular velocity ω/p. This quantity is termed the synchronous angular velocity, denoted by ω_s and measured in radians per second.

ANALYSIS OF MULTI-POLE MACHINES

If $p\theta$ is replaced by θ_e, equations 4.65 and 4.67 take the same form as the equations for a two-pole winding, and the analysis of two-pole machines may be applied directly to multi-pole machines. For the rest of this book a two-pole machine will be assumed unless the contrary is stated. To extend the analysis to multi-pole machines, it is only necessary to multiply the synchronous speed by $1/p$ and the torque by p. The torque multiplication arises from the derivation of the torque equation, in which an integral is taken round the whole circumference of the rotor; since this comprises $2p$ pole pitches, i.e. p complete cycles of the field, the torque equation becomes

$$T = pKB_{1m}B_{2m} \sin \delta_{12}, \tag{4.68}$$

where δ_{12} is the electrical angle between the axes of the stator and rotor fields. The work done per second by the rotating field is $\omega_s T$, and this is independent of the number of poles — as it must be, since the voltage and current determine the input power, and these quantities are unaffected by the number of poles.

4.7 Introduction to practical windings

For most of this chapter ideal sinusoidally distributed windings have been postulated, in order to simplify the mathematical treatment. One consequence of a sinusoidal distribution is that the rotating field has a constant amplitude B_m and a constant angular velocity ω_s. These are desirable properties, which ensure that the torque developed by the machine will be a steady quantity. Practical windings are designed to approximate to this ideal, and in this section we show what can be achieved with a fairly simple arrangement of conductors in the form of coils.

Fig. 4.18 shows a distribution of conductors which gives a stepped approximation to the ideal sinusoid for the α phase, and Fig. 4.19 shows the corresponding m.m.f. diagram. Conductors carrying i_α inwards are designated α, and those carrying i_α outwards are designated $\bar{\alpha}$. Similar diagrams for the β phase are shown

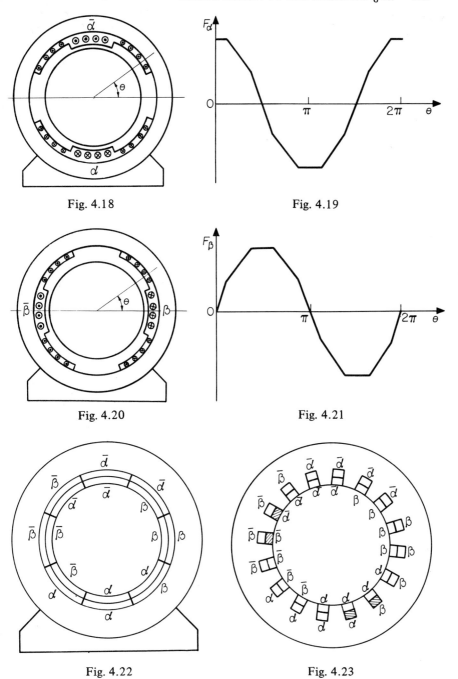

Fig. 4.18

Fig. 4.19

Fig. 4.20

Fig. 4.21

Fig. 4.22

Fig. 4.23

in Figs. 4.20 and 4.21. These two conductor arrangements can be combined, as shown in Fig. 4.22, to form a layer of conductors of uniform thickness. In practice the conductors in a machine are placed in slots, as illustrated in Fig. 4.23. A conductor carrying current inwards in one slot may be joined with another conductor carrying the same current outwards in another slot to form a coil, as shown in Fig. 4.24. Since one coil side is at the top of a slot and the other side at the bottom of a slot, the process may be continued in this way all round the periphery; all conductors are joined in pairs to form coils, and all coils have the same shape. Fig. 4.25 shows coils of this kind fitted into the stator core of a

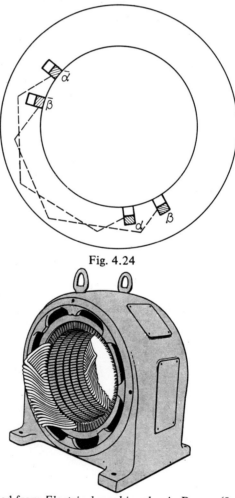

Fig. 4.24

Fig. 4.25 Reproduced from *Electrical machines* by A. Draper (Longmans, 2nd edition)

machine. After winding, the coils are interconnected so that all the conductors of one phase are in series (or possibly in series/parallel groups when the machine has more than two poles). It may be observed that this method of constructing an a.c. stator winding is very similar to the construction of a d.c. armature winding mentioned in section 2.1.

The principle of this two-phase winding can easily be extended to three phases, and it is only one of many possible winding arrangements. The intricacies of practical windings do not concern us here; they are treated in standard texts such as Say [4]. It is sufficient to know that windings can be made to produce an acceptable approximation to the ideal sinusoidal m.m.f. distribution, and for the purposes of analysis we assume that the m.m.f. is exactly sinusoidal.

Problems

4.1 In the definition of the m.m.f. of a distributed winding (section 4.1), the summation of currents is taken from a starting point θ_0 for which $H = 0$. If θ_0 cannot be found by inspection, some arbitrary angle, say $\theta = 0$, must be used as the starting point. Then

$$F(\theta) = F'(\theta) + F(0),$$

where $F'(\theta)$ is the sum of the currents in the conductors between 0 and θ, and $F(0)$ is the m.m.f. at the point $\theta = 0$.

Use the fact that the net flux out of a closed surface is always zero to prove that when the air-gap length g is a constant the m.m.f. $F(\theta)$ must satisfy the condition

$$\int_0^{2\pi} F(\theta)\, d\theta = 0.$$

Hence show how the value of $F(0)$ may be deduced from a graph of $F'(\theta)$ against θ.

4.2 Extend the analysis of a three-phase winding given in section 4.4 to m phases. Show that the winding will produce a magnetic field rotating with an angular velocity equal to the angular frequency of the supply, and that the amplitude will be constant and equal to $m/2$ times the maximum field produced by any one phase acting alone. Note that the analysis involves the summation of terms of the form $\cos(\omega t + \theta + 4\pi r/m)$ where r is an integer. This is most easily accomplished if the terms are represented by phasors.

4.3 Verify the expression given in equation 4.49 for the e.m.f. induced in the α-phase winding by considering an elementary coil formed by the conductors in two elementary arcs of angle $d\theta$, one located at $+\theta$ and the other at $-\theta$. If the conductor density is $k \sin \theta$, calculate the flux linkage of this elementary coil, and by integrating from $\theta = 0$ to $\theta = \pi$ evaluate the total flux linkage of the winding

4.4 The relationship between the m.m.f. and the air-gap flux density is given by the expression

$$B(\theta) = \frac{\mu_0}{g} F(\theta),$$

where g is the radial length of the air gap. Let $P = \mu_0/g$; if the air-gap length varies with θ, then P will be a function of θ and we have

$$B(\theta) = P(\theta) F(\theta).$$

Consider a machine with a circular stator, and a stator winding which produces a rotating m.m.f. given by

$$F = KI_m \cos(\theta - \omega t - \alpha),$$

where K is a constant and I_m is the maximum value of the current in one phase of the stator. The machine has an iron rotor with no windings; it is not circular, but is shaped so that the function P takes the form

$$P = A_0 + A_1 \cos 2\chi,$$

where χ is the angular displacement from a reference axis on the rotor. If the rotor revolves with an angular velocity ω_r, the position of a point on the rotor with respect to the stator is $\theta = \chi + \omega_r t$, and P becomes

$$P = A_0 + A_1 \cos 2(\theta - \omega_r t).$$

Show that the current on the stator between θ and $\theta + d\theta$ is given by

$$di = \frac{\partial F}{\partial \theta} d\theta,$$

and hence find the total torque exerted on the stator winding; this will be equal and opposite to the torque exerted on the rotor. Show that the torque will be constant if $\omega_r = \omega$, and compare this machine with the one considered in problem 1.6.

References

[1] CHAPMAN, C. R. (1965). *Electromechanical energy conversion.* Blaisdell Publishing Co., New York.
[2] FITZGERALD, A. E. KINGSLEY, C., Jr. and KUSKO, A. (1971). *Electric machinery,* 3rd edition. McGraw-Hill, New York.
[3] TAYLOR, P. L. (1964). *Servomechanisms,* 2nd edition. Longmans, London.
[4] SAY, M. G. (1958). *The performance and design of alternating current machines,* 3rd edition. Pitman, London.

CHAPTER 5

Synchronous Machines

The synchronous machine in its normal form consists of a stator with a poly-phase winding (termed the armature winding) which produces a rotating magnetic field, and a magnetized rotor having the same number of poles as the stator field. The rotor may be a permanent magnet, or it may be magnetized by direct current flowing in a winding termed the field or excitation winding. Wound-rotor machines are the most common, for the ability to vary the rotor excitation is an important feature of synchronous machine operation.

As its name implies, the distinctive feature of the synchronous machine is that the rotor revolves in synchronism with the rotating magnetic field of the stator, and its speed is therefore related to the frequency of the a.c. mains supply to the stator. Synchronous machines can operate as generators or motors; nearly all the generators in power supply systems are of this kind, and large synchronous motors are widely used as high-efficiency constant-speed industrial drives.

An important step in the theory of the synchronous machine is the development of an equivalent circuit. This concept has already been used with the transformer, to represent the behaviour of magnetically-coupled coils by a network of ideal circuit elements (resistors, inductors and an ideal transformer). In a similar way the interaction of currents and magnetic fields in the synchronous machine can be represented by a simple circuit made up of ideal elements, and the characteristics of the machine are readily deduced from the circuit. Thus the equivalent circuit forms a link between the internal electromagnetic processes and the external performance characteristics.

5.1 Phasor diagram and equivalent circuit

We assume that the synchronous machine has a cylindrical rotor, so that it is reasonable to postulate sinusoidally distributed magnetic fields. Fig. 5.1 is a space phasor diagram representing the magnetic field components at a particular instant of time, and the phasors rotate with the synchronous angular velocity ω_s. In this diagram, $\vec{B}_1$ is the magnetic field produced by currents in the stator winding; $\vec{B}_2$ is the magnetic field produced by currents in the rotor winding; and $\vec{B}$ is the total or resultant magnetic field. The component fields, acting individually, would induce voltages E_1 and E_2 in one phase of the stator armature winding. The total induced voltage E_t is thus equal to $E_1 + E_2$, as shown in the time phasor diagram of Fig. 5.2.

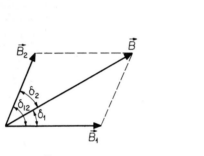

Fig. 5.1

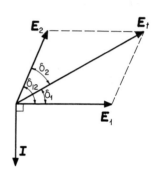

Fig. 5.2

If resistance and leakage reactance may be ignored, the total induced voltage E_t will be equal to the terminal voltage V. From equation 4.51, the component voltage E_1 (due to the stator magnetic field) is related to the stator current by

$$E_1 = j\omega L_m I = jX_m I, \tag{5.1}$$

where L_m is the effective inductance of one stator phase, and X_m is the corresponding reactance. The component E_2 is the contribution of the rotor magnetic field to the total induced voltage; since it depends on the rotor excitation current, it is termed the excitation voltage, usually denoted by E. With these changes of nomenclature the phasor diagram of Fig. 5.2 may be re-drawn, taking $V = E_t$ as the reference phasor; the result is shown in Fig. 5.3, where the angle δ is the same as δ_2 in Fig. 5.2.

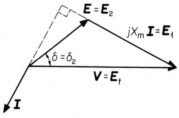

Fig. 5.3

COMPLETE EQUIVALENT CIRCUIT

Fig. 5.3 is the phasor diagram for one stator phase of a synchronous machine; by inspection it is also the phasor diagram for the circuit shown in Fig. 5.4, which is therefore an *equivalent circuit* of one phase of the machine. The resistance and leakage reactance of the armature winding may be included as elements in series with the reactance X_m, and the stator core loss may be represented by a shunt resistance R_l, to form the complete equivalent circuit shown in Fig. 5.5.

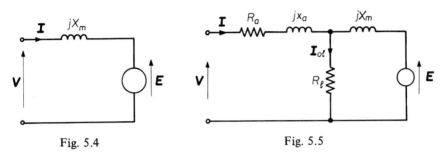

Fig. 5.4   Fig. 5.5

It is useful to transform the complete equivalent circuit so that the reactance X_m appears as a shunt element; this permits a direct comparison with the equivalent circuit of the induction machine which is derived in chapter 6. The transformation is accomplished by replacing the series combination of E and jX_m with an equivalent parallel circuit, as shown in Fig. 5.6. The complete equivalent circuit for the synchronous machine then takes the form shown in Fig. 5.7, and

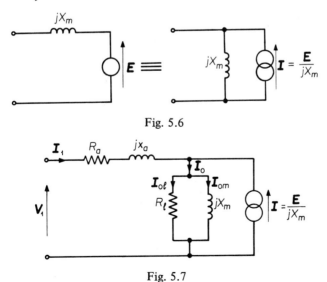

Fig. 5.6

Fig. 5.7

the resemblance to the equivalent circuit of the transformer will be noted. In this circuit (Fig. 5.7) the reactance X_m is a mutual or magnetizing reactance, with the magnetizing current I_{0m} supplied partly by the stator current and partly by the rotor excitation. The flux associated with this reactance represents the total magnetic field in the machine air gap, which is the resultant of fields produced by the stator and rotor currents.

APPROXIMATE EQUIVALENT CIRCUIT

The armature resistance R_a is often small in comparison with the reactance X_m, and it may be a reasonable approximation to ignore it. If the core loss resistance R_l is also ignored, the two reactances x_a and X_m in Fig. 5.5 may be combined to form a single reactance X_s, known as the synchronous reactance. The equivalent circuit then takes the simple form shown in Fig. 5.8, which is

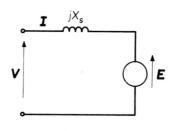

Fig. 5.8

quite a good representation of large machines. With small machines the resistance R_a can be more than 10% of the synchronous reactance X_s, and it should be included in series with X_s if greater accuracy is required. The omission of the core loss resistance R_l has very little effect for most conditions of operation, since the current I_{0l} (Fig. 5.5) is very small in comparison with the normal value of the total current I.

The phasor equation for the equivalent circuit of Fig. 5.8 is

$$V = E + jX_s I, \tag{5.2}$$

and when the resistance R_a is included this becomes

$$V = E + (R_a + jX_s) I. \tag{5.3}$$

The representation of equation 5.2 by a phasor diagram will be considered in the next section. Equations 5.2 and 5.3 provide an interpretation of the excitation voltage E; when $I = 0$, $V = E$, and E is therefore the open-circuit terminal voltage of the machine.

5.2 Synchronous machine characteristics

The essential features of synchronous machine operation may be derived from the approximate equivalent circuit and the corresponding phasor diagram; before doing so, however, it is useful to make some general deductions from the rotating field concepts.

SYNCHRONOUS SPEED

It was shown in section 4.6 that the angular velocity of the rotating magnetic field is given by $\omega_s = \omega/p$, where ω is the angular frequency of the supply and p is the number of pole pairs. If the frequency of the mains supply is f hertz, it follows that the synchronous speed of the machine will be f/p rev/s, or $60f/p$ rev/min. With a mains frequency of 50 Hz the speed of a two-pole machine will be 3 000 rev/min, the speed of a four-pole machine will be 1 500 rev/min, and so on. Once the number of poles has been selected by the designer, the speed of a synchronous machine can only be altered by varying the supply frequency. Since this is not usually practicable, the synchronous machine is a constant speed device.

SYNCHRONOUS TORQUE

When the rotor is running at the synchronous speed, its poles will be displaced by a constant angle from the effective poles of the stator. The magnetic lines of force at a particular instant of time are as shown in Fig. 4.11, and this magnetic field pattern rotates without changing its form. A constant electromagnetic torque is therefore exerted on the rotor; and this will balance the mechanical torque applied to the shaft. The torque equation deduced for stationary fields may therefore be applied to the synchronous machine, giving

$$T = KB_{2m}B_m \sin \delta_2. \qquad [4.23]$$

Now $B_m \propto V$, $B_{2m} \propto E$ and $\delta_2 = \delta$; the torque equation becomes

$$T \propto VE \sin \delta, \qquad (5.4)$$

and if the terminal voltage V and excitation voltage E are held constant then $T \propto \sin \delta$.

Load angle

Adopting the convention that positive values of T and δ corresponding to motoring operation, the variation of torque with the angle δ is as shown in Fig. 5.9. This is an important characteristic of the synchronous machine, and the angle δ is termed the *load angle*; its value varies according to the load applied to the shaft, within the limits of $\pm \pi/2$ radians. Since the speed of the machine is fixed, the torque is entirely determined by the mechanical system connected to the machine shaft. If this is a mechanical load, the torque T will be positive; the synchronous machine acts as a motor, and the positive value of δ implies that

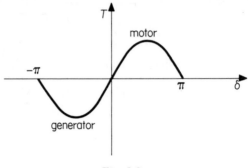

Fig. 5.9

the stator poles are ahead of the rotor poles. If a source of mechanical power is coupled to the shaft, T will be negative and the synchronous machine will act as a generator. The angle δ will also be negative, showing that the stator poles are now lagging behind the rotor poles. The magnitude of the torque has a maximum value, known as the pullout torque, when $\delta = \pm\pi/2$ radians. If a torque in excess of this value is applied to the machine shaft, the electromagnetic torque cannot balance the shaft torque, and synchronism will be lost.

SYNCHRONOUS MOTORS AND GENERATORS

As with the d.c. machine, there is no essential difference between motoring and generating operation of the synchronous machine. Fig. 5.10 shows the phasor diagram when the machine is operating as a motor; the terminal voltage

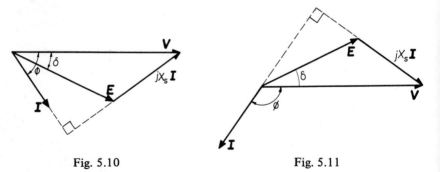

Fig. 5.10   Fig. 5.11

V leads the excitation voltage E by the load angle δ, and the armature current I therefore has a component in phase with V. When the shaft torque is reversed, so that the machine is driven as a generator, the phasor diagram takes the form shown in Fig. 5.11. The voltage V lags E by the load angle δ, and the current I has a component in antiphase with V.

Suppose that the shaft torque gradually changes from a positive (motoring) value to a negative (generating) value, with the values of E and V held constant; the phasor diagram will gradually change from the form shown in Fig. 5.10 to

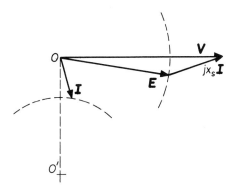

Fig. 5.12

the one shown in Fig. 5.11. The locus of E will be an arc of a circle with its centre at 0, as shown in Fig. 5.12, and the locus of I will be another arc with its centre at $0'$. When the torque is zero the load angle δ will be zero; I will be in quadrature with V, and the electrical power will also be zero.

SYNCHRONOUS MOTOR CHARACTERISTICS

It has already been mentioned that the ability to vary the rotor (or field) excitation is an important feature of the synchronous machine, and we now consider the effect of such a variation when the machine operates as a motor with a constant load. Similar results hold for the synchronous generator with constant mechanical input power.

When the torque load on the motor is constant the power output will be constant, and if losses are neglected there will be a constant input power per phase given by

$$P = VI \cos \phi. \tag{5.5}$$

If the voltage V is constant this equation implies that $I \cos \phi$ is constant, and the locus of the current phasor I is the line AB in Fig. 5.13. From this diagram

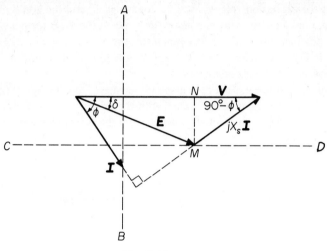

Fig. 5.13

we have

$$MN = E \sin \delta = X_s I \sin(90 - \phi),$$

i.e. $$E \sin \delta = X_s I \cos \phi; \qquad (5.6)$$

thus $E \sin \delta$ is a constant, and the locus of E is the line CD. If $\phi = 0$ when $E = E_0$, the machine will operate at unity power factor and I has a minimum value. When $E < E_0$, ϕ is negative and the machine takes a lagging current, as shown in

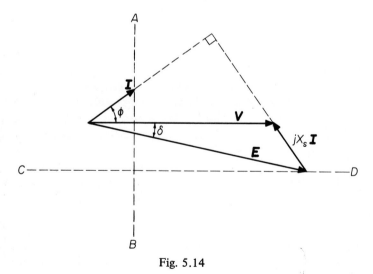

Fig. 5.14

Fig. 5.13; the machine is said to be under-excited, and synchronism will be lost when $\delta = \pi/2$. When $E > E_0$, the machine is over-excited; ϕ is positive and the machine takes a leading current (Fig. 5.14). If the current I is plotted against the excitation voltage E for different values of the power P, the result is a set of curves known as V-curves (Fig. 5.15). A useful characteristic of the synchronous motor is the leading phase angle of the current when the machine is over-excited. It can be used to compensate for the lagging current taken by other loads such as induction motors, so that the total load power factor is unity. Thus, in Fig. 5.16, if the lagging load current is I_l and the synchronous motor current I_s is suitably adjusted by controlling the excitation, the total current I will be in

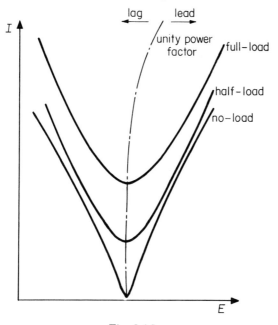

Fig. 5.15

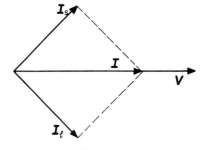

Fig. 5.16

phase with the voltage V. This is known as power factor correction, and the synchronous machine can be used solely for this purpose, with no mechanical load connected. Then

$$VI \cos \phi = P \to 0; \tag{5.7}$$

therefore $I \cos \phi \to 0$, and the phasor diagram (Fig. 5.14) shows that $\phi \to \pi/2$. Since the current leads the voltage by approximately $\pi/2$ radians, the machine is known as a synchronous capacitor; it is also known as a synchronous compensator.

Torque equation

The torque equation for the synchronous motor may be deduced from the phasor diagram. Let $\omega_s = \omega/p$ be the synchronous angular velocity; the total mechanical output power is $\omega_s T$, and if there are m phases the power per phase is $\omega_s T/m$. If losses are neglected this must be equal to the electrical input power per phase:

$$\frac{\omega_s T}{m} = P = VI \cos \phi. \tag{5.8}$$

Substituting for $\cos \phi$ from equation 5.5 gives

$$T = \frac{m}{\omega_s X_s} VE \sin \delta. \tag{5.9}$$

This agrees with equation 5.4, which was derived from a consideration of the magnetic fields.

STARTING OF SYNCHRONOUS MOTORS

When the rotor is running at less than the synchronous speed, the load angle δ will increase continuously. Fig. 5.9 shows that the torque T will be alternately positive and negative, with a mean value of zero. A synchronous motor is therefore not inherently self-starting, and there must be some means of accelerating the rotor to a speed close to synchronism. If the inertia of the rotor (together with any coupled load) is low enough, the first positive half-cycle of torque will then pull the rotor into step with the field. Induction motors develop a positive torque at speeds down to zero, and the induction motor principle is generally used for starting synchronous machines. Occasionally a small induction motor (known as a 'pony motor') is coupled to the synchronous machine shaft; this motor is connected to the a.c. supply for the starting process, and disconnected once the main machine rotor is synchronized. More commonly, the synchronous machine rotor is provided with conducting paths for induced currents, so that the resulting induction torque will accelerate the rotor. These conducting paths may take the form of a separate winding on the rotor, designed in accordance with induction motor principles; or the rotor may have solid steel poles instead of a laminated structure, so that induced currents flow in the material of the rotor core. In all methods of starting, the rotor is left unexcited until it is near to the synchronous speed; application of the excitation current then causes the rotor to 'pull into step' or 'synchronize'.

5.3 Salient-pole synchronous machines

The theory developed in this chapter is only valid for machines with a uniform air gap, i.e. those in which the rotor and the stator bore are both cylindrical. When the rotor has salient poles (as shown in Fig. 5.17) it is no longer possible to represent the machine by a simple equivalent circuit, and the torque equation

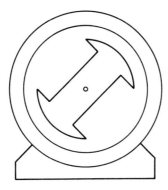

Fig. 5.17

contains an additional term. An outline of the theory is given in section 7.4, and the torque equation developed there for a two-pole, two-phase machine may be generalized for $2p$ poles and m phases:

$$T = \frac{mp}{2\omega}\left[\frac{2VE}{X_d}\sin\delta + V^2\left\{\frac{1}{X_q} - \frac{1}{X_d}\right\}\sin 2\delta\right]. \qquad (5.10)$$

$$\qquad\quad\text{(a)}\qquad\qquad\qquad\text{(b)}$$

Term (a) in this equation represents the normal synchronous torque, which may be identified with equation 5.9. Term (b) represents a component of torque due to the saliency of the rotor; this component is termed the reluctance torque, and it vanishes when the rotor is cylindrical, for X_d is then equal to X_q. The meanings of X_d and X_q, and the reason for the name reluctance torque, are explained in section 7.4.

A qualitative explanation of the origin of the reluctance torque is that the salient poles of the rotor will tend to line up with the axes of the stator magnetic field. The torque will be zero when the rotor pole axis is in line with the stator field axis, and zero again when the rotor pole axis is at right angles to the stator field axis. In contrast, the synchronous torque will have a maximum or a minimum value when the two axes are at right angles. The reason for the difference is that the reluctance torque is dependent on the induced magnetization of the rotor, which varies with the rotor position; whereas the synchronous torque depends on the rotor magnetization produced by the field winding, and this is

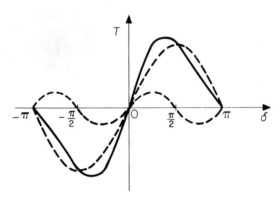

Fig. 5.18

independent of the rotor position. Thus in equation 5.10 the reluctance torque varies as $\sin 2\delta$, while the synchronous torque varies as $\sin \delta$. The torque/load-angle characteristic for a salient-pole machine therefore takes the form shown in Fig. 5.18.

SYNCHRONOUS RELUCTANCE MOTORS

An interesting possibility is to omit the rotor field winding from a salient-pole synchronous motor, so that only the reluctance torque term remains in equation 5.10. Machines of this kind are known as synchronous reluctance motors, and they normally incorporate on the rotor an induction starting winding of the cage type (see chapter 6). Reluctance motors combine the advantages of the cage induction motor with the speed characteristic of the synchronous machine. They are used in place of induction motors when a constant speed, related to the supply frequency, is required. A typical application is to maintain an absolute speed relationship between a number of widely separated shafts, where mechanical coupling would be difficult.

Problems

5.1 A three-phase star-connected synchronous generator has a reactance of $10\,\Omega$ per phase, and it operates with a constant line voltage of $520\,\text{V}$. When the generator is delivering its normal rated power the line current is 40 A and the power factor is unity. Calculate the output power and the magnitude of the excitation voltage per phase under these conditions.

With the excitation voltage unchanged, the output power of the generator is increased to its maximum value. Calculate the new values of line current, power factor and output power.

5.2 In the steady state a synchronous motor operates with a load angle δ_0 and it delivers a torque T_0 to the load. A momentary fall in the supply voltage causes the load angle to increase by a small amount, and after the voltage disturbance has passed the load angle is given by

$$\delta = \delta_0 + \epsilon.$$

If the electromagnetic torque is then given by

$$T = T_0 + \Delta T,$$

show that

$$\Delta T = \frac{T_0}{\tan \delta_0}\,\epsilon.$$

It may be assumed that the machine has a cylindrical rotor; that the steady-state equations are applicable; and that the armature resistance may be neglected.

5.3 If the torque load on the motor in problem 5.2 remains unchanged, show that the rotor equation of motion is

$$\frac{J}{p}\frac{d^2 \epsilon}{dt^2} + \frac{T_0}{\tan \delta_0}\,\epsilon = 0,$$

where J is the moment of inertia of the rotor and p is the number of pole pairs. Hence show that there will be small oscillations superimposed on the steady motion of the rotor, and find the frequency of the oscillations.

5.4 With the machine considered in problem 5.3, show that there will be an alternating voltage induced in the rotor field winding, and explain what effect this might be expected to have on the motion of the rotor.

5.5 In chapter 7 it is shown that the general equations for a synchronous machine reduce to the following form when the rotor is cylindrical and the armature resistance is neglected:

$$V \sin \delta = X_s I \cos(\phi - \delta),$$

$$V \cos \delta = E + X_s I \sin(\phi - \delta).$$

Show that these equations are satisfied by the phasor diagram of Fig. 5.10.

CHAPTER 6

Induction Machines

An essential feature of the operation of the synchronous machine is that the rotor runs at the same speed as the rotating magnetic field produced by the stator winding; the magnetic field as 'seen' from a point on the rotor does not vary with time. A very different type of machine results if the rotor is allowed to run more slowly than the rotating field; the rotor will 'see' a rotating field moving past it at the difference of the two speeds, and this will cause induced currents to flow in suitably arranged conductors on the rotor. These currents will interact with the rotating field to produce a torque, and this is the basis of the induction motor.

In common with other rotating machines, induction machines can operate as motors or generators. For reasons that will be discussed later induction generators have a very limited use, and nearly all electric power is generated by synchronous machines. Induction motors, on the other hand, are used in far greater numbers than any other type of machine; they range in power rating from a few watts to tens of megawatts. The simplicity of the induction principle is reflected in the robust, reliable and relatively inexpensive construction of the machine itself, and the induction machine is the natural choice in the majority of motor applications.

A proper understanding of the behaviour of induction machines must include the rotating field concepts outlined in section 6.1 and the equivalent circuit developed in section 6.2. These provide the basis for the derivation of the steady-state characteristics in section 6.3, and for the discussion of speed control and single-phase operation in the remainder of the chapter.

6.1 Electromagnetic action

The stator (or 'primary') winding of an induction machine is similar to the stator of a synchronous machine, and when connected to a suitable a.c. supply it will produce a rotating magnetic field of the form given by equation 4.37:

$$B_1 = B_{1m} \cos(\omega t - \theta). \tag{6.1}$$

This equation implies that the machine has only two poles, and for simplicity the analysis in this section and in section 6.2 will be given for a two-pole machine. The results are extended at the beginning of section 6.3 to the general case of a $2p$-pole machine.

Suppose that the rotor of the induction machine rotates with an angular velocity ω_r, so that at time t a reference axis on the rotor makes an angle given by

$$\psi = \omega_r t \tag{6.2}$$

with the reference axis of the stator (Fig. 6.1). Let χ be the angular position of a

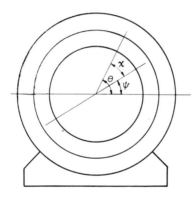

Fig. 6.1

point on the rotor, measured from the rotor reference axis; then the angle measured from the stator reference axis is

$$\theta = \chi + \psi = \chi + \omega_r t. \tag{6.3}$$

Substitution of this expression for θ in equation 6.1 gives the magnetic field in terms of the angle χ:

$$B_1 = B_{1m} \cos\{(\omega - \omega_r)t - \chi\}. \tag{6.4}$$

This equation implies that the rotor 'sees' a magnetic field rotating past it with an angular velocity of $\omega - \omega_r$, known as the slip angular velocity. The quantity

$$s = \frac{\omega - \omega_r}{\omega} \tag{6.5}$$

is termed the fractional slip, and we may re-write equation 6.4 in the form

$$B_1 = B_{1m} \cos(s\omega t - \chi). \tag{6.6}$$

This rotating field will induce an e.m.f. at the slip angular frequency $s\omega$ in any coils on the rotor.

ROTOR MAGNETIC FIELD

If the rotor carries a polyphase winding similar to the stator winding, a polyphase system of voltages will be generated by the rotating field; and if the terminals are connected to a balanced resistive load, balanced polyphase currents will flow at the slip frequency $s\omega$. These currents, flowing in the rotor (or 'secondary') winding, will set up a rotating m.m.f. wave of the form

$$F_2 = F_{2m} \cos(s\omega t - \chi - \delta_{12}) \tag{6.7}$$

where δ_{12} is a constant but as yet unknown angle.

The rotor m.m.f. will in turn set up a rotor magnetic field B_2, which has the same form as the m.m.f. wave:

$$B_2 = B_{2m} \cos(s\omega t - \chi - \delta_{12}). \tag{6.8}$$

The total magnetic field is thus $B = B_1 + B_2$, and this may be written as

$$B = B_m \cos(s\omega t - \chi - \delta_1). \tag{6.9}$$

It is this total field B which determines the e.m.f. induced in the rotor winding when rotor currents are flowing.

ROTATING FIELD CONCEPTS

The fields B_1 and B_2 rotate with the same angular velocity; with respect to the rotor, this is the slip angular velocity $s\omega$; with respect to the stator it is the synchronous speed ω. Since there is a constant angle δ_{12} between the axes of the fields, there will be a constant torque on the rotor given by

$$T = KB_{1m}B_{2m} \sin \delta_{12}. \tag{4.24}$$

The mechanism of torque production is thus the same as in the synchronous machine, but the principle of operation is very different. In the synchronous machine, the rotor magnetic field is set up by externally impressed currents, and its axis is fixed relative to the rotor material. The rotor magnetic field of an induction machine, on the other hand, is produced by induced currents in the rotor, and its axis rotates relative to the rotor material. This rotation, or slipping, of the field past the rotor is an essential feature of induction motor operation; if the rotor ran at the synchronous speed it would 'see' a steady field, and there would be no induced currents.

6.2 Equivalent circuit

In order to predict the performance characteristics, it is necessary to construct an equivalent circuit which will represent the machine in terms of lumped circuit parameters and the stator terminal voltage and current. This may be done by considering the relationships between the magnetic field components, the currents flowing in the stator and rotor windings, and the e.m.f.s induced in those windings. The fact that the rotor currents are induced from the stator makes the derivation of the equivalent circuit more difficult than the corresponding derivation for the synchronous machine.

EQUIVALENT CIRCUIT OF AN IDEAL MACHINE

Stator and rotor induced voltages

First consider a machine with the rotor circuits open, so that no rotor currents can flow although there will be induced voltages in these circuits. The rotor magnetic field will be zero, and the total field B will be equal to the stator field B_1. This rotating field will induce an e.m.f. in each phase of the stator winding, and from equation 4.49 the amplitude of the e.m.f. is given by

$$E'_m = \pi r l k_1 \omega B_m. \tag{6.10}$$

If the stator winding has negligible resistance and leakage reactance, the induced e.m.f. E' will be equal to the terminal voltage V_1. From equation 4.49, the current required to set up the field is given by

$$I_1 = \frac{V_1}{j\omega L_m} = \frac{V_1}{jX_m}, \tag{6.11}$$

where L_m is the effective inductance of one phase of the winding, and X_m is the corresponding reactance (known as the magnetizing reactance). Thus with the

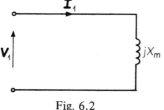

Fig. 6.2

rotor on open circuit, each phase of the machine may be represented by the equivalent circuit of Fig. 6.2.

The total field B rotates past the rotor with the slip angular velocity $s\omega$. It will induce an e.m.f. of angular frequency $s\omega$ in each phase of the rotor, and from equation 4.47 the amplitude of the e.m.f. will be

$$E_m = \pi r l k_2 s \omega B_m. \qquad (6.12)$$

By comparison with equation 6.10 this amplitude may be written as

$$E_m = s \frac{k_2}{k_1} E'_m. \qquad (6.13)$$

Since k_1 and k_2 are measures of the conductor density, the ratio k_2/k_1 is equal to the turns ratio N_2/N_1 for each phase of the rotor and stator windings. Thus the machine behaves as a kind of transformer, in which the magnitude and frequency of the rotor (or secondary) voltage are proportional to the fractional slip s. For the present, we assume that $k_1 = k_2$, so that $E_m = sE'_m$.

Stator and rotor currents

Now let the rotor circuits be closed, so that a current of magnitude I_2 flows in each phase; this current will alternate with the slip angular frequency $s\omega$. The rotor currents will set up a magnetic field B_2, which combines with the stator field B_1 to give the total field B. We have seen that the total field B determines the magnitude of the voltage induced in each stator phase, and this in turn is equal to the applied voltage V_1. Thus the value of B_m must remain constant, and the presence of the rotor field B_2 implies that the stator field B_1 must change, and with it the stator current I_1. We may regard the current I_1 in one phase of the stator as the sum of two components: a constant magnetizing component I_{0m}, which sets up the constant net rotating field B; and a load component I'_2, which produces a magnetic field B'_2 equal and opposite to the rotor field B_2. Since the stator and rotor windings are similar, the magnitude of I'_2 must be equal to the magnitude of the current I_2 in one rotor phase, although the frequencies will be different.

Relationship between voltages and currents

The magnitude of the e.m.f. induced in one rotor phase is $E = sE'$, and this is impressed on a circuit consisting of the rotor leakage inductance l_2 in series with a resistance R_2 (which represents the winding resistance plus any external resistance). Since the rotor angular frequency is $s\omega$, the magnitude of the rotor current and its phase relationship with the induced e.m.f. is given by the phasor equation

$$I_2 = \frac{E}{R_2 + js\omega l_2}. \qquad (6.14)$$

We wish to find the magnitude of the stator current I_2', and its phase relationship with the stator induced e.m.f. E'. Because of the frequency difference we cannot directly equate stator and rotor phasor quantities, and the derivation takes the following form.

The stator magnetic field component B_2' opposes the rotor field B_2, and apart from the change of sign their axes must make the same angle with the axis of the total field B. The induced e.m.f.s in the windings are related to B, while the currents I_2' and I_2 are related to the fields B_2' and B_2 respectively. It follows that the angle between E' and I_2' for one phase of the stator winding is equal to the angle between E and I_2 for one phase of the rotor winding. Let I_2' be the current which flows when the impedance Z' is connected to the e.m.f. E'. Then

$$I_2' = \frac{E'}{Z'}. \tag{6.15}$$

Since $I_2 = I_2'$ and $E = sE'$, equations 6.14 and 6.15 give the relationship

$$\frac{E'}{Z'} = \frac{sE}{|R_2 + js\omega l_2|},$$

i.e.

$$Z' = \frac{1}{s}|R_2 + js\omega l_2|. \tag{6.16}$$

Since the phase angles are equal, we must also have

$$arg\,Z' = arg(R_2 + js\omega l_2). \tag{6.17}$$

Equations 6.16 and 6.17 are both satisfied if

$$Z' = \frac{1}{s}(R_2 + js\omega l_2)$$

$$= \frac{R_2}{s} + j\omega l_2 \tag{6.18}$$

$$= \frac{R_2}{s} + jx_2,$$

where $x_2 = \omega l_2$ is the rotor leakage reactance at the stator supply frequency ω.

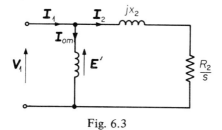

Fig. 6.3

One stator phase of the induction machine may therefore be represented by the equivalent circuit shown in Fig. 6.3. The elements x_2 and R_2/s represent the effects of the rotor (or secondary) in the stator (or primary) circuit.

COMPLETE EQUIVALENT CIRCUIT

The similarity of the induction motor to a transformer with a closed secondary circuit should be noted. The relative motion between the primary and secondary is represented in the equivalent circuit by the factor $1/s$ multiplying the secondary resistance R_2, and the significance of this term will be explained later. By analogy with the transformer, the effects of stator resistance, stator leakage reactance and core loss may be included in the equivalent circuit. Fig. 6.4 shows the complete equivalent circuit of the machine; in this circuit the

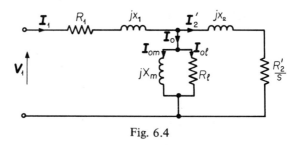

Fig. 6.4

effect of a turns ratio $1:n$ between the stator and rotor windings is also included, by using referred values of the secondary parameters ($R_2' = R_2/n^2$; $x_2' = x_2/n^2$).

It is instructive to compare the equivalent circuit of the induction machine with the complete equivalent circuit of the synchronous machine shown in Fig. 5.7. In the synchronous machine equivalent circuit, the rotor is represented by an active element — a current generator — which is capable of supplying some or all of the magnetizing current I_{0m}. In the induction machine, on the other hand, there is no external rotor excitation; the rotor is represented by an impedance element in the equivalent circuit, and all the magnetizing current must be drawn from the stator supply. This means that an induction motor necessarily behaves as an inductive load, taking current at a lagging power factor.

APPROXIMATE EQUIVALENT CIRCUIT

The equivalent circuit shown in Fig. 6.4 is a fairly accurate representation of the machine, and it may be used to predict the characteristics. The analysis is greatly simplified, however, if two approximations are made. First, the shunt elements R_l and X_m are transferred to the input terminals. This was shown to be a good approximation with a power transformer (see section 3.3); it is less

satisfactory with the induction machine, because the air gap between the stator
and rotor reduces the value of X_m and increases x_1, in comparison with a
transformer of similar rating. The second approximation is to ignore the stator
resistance R_1 in comparison with the term R_2'/s. This is a good approximation
when the slip is small, and the equivalent circuit then takes the form shown in
Fig. 6.5; the primary and secondary leakage reactances are combined to give a

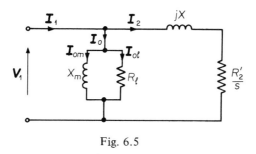

Fig. 6.5

total leakage reactance $X = x_1 + x_2'$. This simplified circuit will demonstrate the
essential features of induction motor performance, and the complete equivalent
circuit can always be used when a more accurate calculation is required.

6.3 Induction machine characteristics

In developing the theory we have assumed that the rotor of an induction
motor carries an m-phase winding of similar construction to the stator winding.
Connection to the rotor is made via brushes and sliprings, and this kind of
machine is known as a wound-rotor or slipring induction machine. More
commonly, the rotor winding consists of a number of bars (usually of aluminium
or a copper alloy) which pass through slots in the laminated iron core of the
rotor. The bars are connected to conducting rings (known as end-rings) at each
end of the core, so that any pair of conductors forms a single short-circuited
turn. From the construction of the rotor, this is termed a cage induction
machine. If the bars are imagined to be divided into m groups, they effectively
form an m-phase winding with each phase short-circuited. Thus the same theory
applies, and the resistance R_2' in the equivalent circuit represents the effective
resistance of each 'phase' of the rotor cage. The value of R_2' can be controlled at
the design stage by adjusting the dimensions of the bars and the resistivity of the
material. Cage rotors are cheap and robust — in small sizes the bars and end-rings
are die-cast in a single operation. The more expensive wound-rotor construction
is only used when it is necessary to control the characteristics of the machine by
varying the external resistance connected across each rotor phase. The construction of typical rotors is shown in Fig. 6.6.

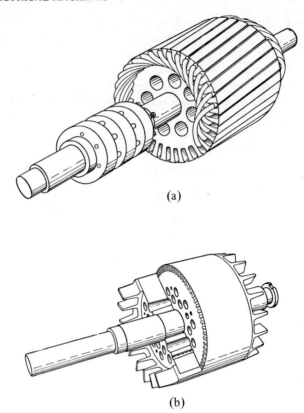

(a)

(b)

Fig. 6.6. (a) Wound rotor, (b) cage rotor. Reproduced from *Electrical machines* by A. Draper (Longmans, 2nd edition).

MULTI-POLE MACHINES

If the machine stator winding has p pairs of poles, the synchronous speed is $\omega_s = \omega/p$. The fractional slip is now defined as

$$s = \frac{\omega_s - \omega_r}{\omega_s}, \qquad (6.19)$$

and the slip angular velocity is $\omega_s - \omega_r = s\omega_s$. Since the rotor winding also has p pairs of poles, the angular frequency of the rotor currents will be p times the slip angular velocity, i.e. $ps\omega_s = s\omega$. Thus the rotor frequency is unchanged, and the same equivalent circuit holds for a machine with any number of pole pairs.

ROTOR POWER RELATIONSHIPS

In the equivalent circuit of Fig. 6.4, the power loss in R_1 and R_l represents the primary copper loss and the core loss. The power loss in the resistance R_2'/s must therefore represent the average input of power to the rotor, for there can be no dissipation of power in the reactances X_m and x_2'. Thus the input of power per phase to the rotor is $(I_2')^2 R_2'/s$; but the power dissipated in the actual resistance of the rotor circuit is only $(I_2')^2 R_2'$. The difference between these quantities is

$$(I_2')^2 R_2' \frac{1-s}{s}$$

and this must represent electrical energy converted into mechanical form. If P is the total power absorbed by the rotor, then

electromagnetic power input to rotor $= m(I_2')^2 \dfrac{R_2'}{s} = P,$ (6.20)

power loss in rotor resistance $= m(I_2')^2 R_2' = sP,$ (6.21)

mechanical power output $= m(I_2')^2 R_2' \dfrac{1-s}{s} = (1-s)P,$ (6.22)

where m is the number of phases. Now consider the torque T exerted on the rotor by the rotating magnetic field. If there are p pairs of poles, the angular velocity of the field is $\omega_s = \omega/p$, and the rotating field therefore does work at the rate $\omega_s T$. This is obviously true if the rotating field is produced by physical poles on the stator, driven mechanically at a speed ω_s; the electromagnetic field is the same when a polyphase winding produces the rotating field, so the work done must be the same. Since the rotor runs at a speed $\omega_r = (1-s)\omega_s$, the mechanical power output is $\omega_r T = (1-s)\omega_s T$. The difference between the work done by the field and the mechanical output must be absorbed in rotor losses, so this is $(\omega_s - \omega_r)T = s\omega_s T$. Thus:

electromagnetic power input to rotor $= \omega_s T,$ (6.23)

power loss in rotor resistance $= (\omega_s - \omega_r)T = s\omega_s T,$ (6.24)

mechanical power output $= \omega_r T = (1-s)\omega_s T.$ (6.25)

These three quantities are in the same ratio as before. Note that the fraction of the input power lost in rotor resistance is equal to the fractional slip s; since there must always be some slip between the rotor and the rotating magnetic field, this represents an unavoidable source of power loss.

The ratio of mechanical power output to electromagnetic power input may be termed the rotor efficiency, and its value is $1 - s$. Since there will be other sources of power loss in the machine the overall efficiency will always be lower than the rotor efficiency, and it is desirable to keep the slip as small as possible.

TORQUE–SPEED CHARACTERISTICS

The torque may be calculated for a given value of slip by equating expressions 6.20 and 6.23 for electromagnetic power, and obtaining the value of I_2' from the equivalent circuit of Fig. 6.5. For a machine with m phases and p pole pairs, we have

$$\omega_s T = \frac{\omega}{p} T = m(I_2')^2 \frac{R_2'}{s} = m \frac{R_2'}{s} \frac{V_1^2}{X^2 + (R_2'/s)^2}. \qquad (6.26)$$

The torque is therefore given by

$$T = \frac{mp}{\omega} \cdot \frac{R_2'}{s} \cdot \frac{V_1^2}{X^2 + (R_2'/s)^2}, \qquad (6.27)$$

which may be written in the alternative form

$$T = \frac{mp}{\omega} \cdot \frac{V_1^2}{X} \cdot \frac{1}{sX/R_2' + R_2'/sX}. \qquad (6.28)$$

Since the slip s is related to the rotor speed ω_r by equation 6.19, equation 6.28 gives the torque–speed relationship for the induction machine. Fig. 6.7 shows a typical torque–speed or torque–slip characteristic. Note that there can be no induced rotor currents when $s = 0$ and $\omega_r = \omega_s$, so the torque must be zero at this point. There are three distinct regions to the torque–speed characteristic shown in Fig. 6.7, which will be considered in turn.

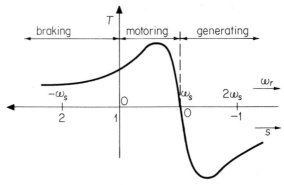

Fig. 6.7

Motoring region

In this region the rotor speed ω_r is positive but less than the synchronous speed ω_s; the torque is also positive, and the machine converts electrical power into mechanical power. The value of the slip s varies from 1 when the rotor is stationary to 0 when the rotor runs at the synchronous speed. In the complete equivalent circuit (Fig. 6.4) the resistance R_2'/s is positive and greater than R_2'; thus the total electrical power absorbed by the rotor exceeds the power dissipated in the rotor copper loss, and the balance is extracted as mechanical power at the shaft. Since the overall efficiency of the machine cannot exceed the rotor efficiency of $1 - s$, induction motors normally operate with a small value of slip. The full-load slip can be as low as 1% in large machines and seldom exceeds 5% in small machines, so the normal rotor speed is always close to the synchronous speed.

Generating region

When the rotor is driven mechanically so that its speed exceeds the synchronous speed, the torque reverses and the machine absorbs mechanical power. The slip is negative in this region, and the resistance R_2'/s is also negative; the rotor therefore exports electrical power to the stator, and the machine acts as a generator. The machine must normally be connected to an a.c. supply before it will act as a generator, since a source of reactive power is required for the magnetizing current flowing in X_m which cannot be provided by the rotor. The synchronous machine provides its own magnetizing current by having a field winding on the rotor, and this is the preferred type of a.c. generator. Induction generators are sometimes used when it is not possible to obtain the constant-speed drive required by a synchronous generator.

Braking region

If the rotor is driven in the opposite direction to the rotating field, the torque will oppose the motion and the machine will again absorb mechanical power. It will not, however, act as a generator. The slip is now greater than 1, and the resistance R_2'/s is positive and smaller than R_2'; the electrical power input to the rotor is insufficient to supply the rotor copper loss, and the balance is supplied by the mechanical power input. The machine therefore acts as a brake, with both the electrical and the mechanical power inputs dissipated in the rotor resistance.

Operation in the braking region can only take place for short periods on account of rotor heating, and it is sometimes used as a method of rapidly stopping an induction motor by 'plugging'. Two of the connections to the three-phase stator are interchanged, thus reversing the direction of rotation of the magnetic field and applying a braking torque to the rotor. The stator supply must be disconnected as soon as the rotor comes to rest, otherwise the rotor will continue to accelerate in the new direction of the rotating field.

INDUCTION MOTOR TORQUE

Several interesting properties of the torque characteristic may be deduced from equations 6.27 and 6.28. When the slip is such that $sX/R'_2 = 1$, the torque will have a maximum value given by

$$T_m = \frac{mpV_1^2}{2\omega X}.$$ (6.29)

This is known as the breakdown torque, and a load in excess of this value will stop the motor. The normal operating region lies to the right of the peak, and the normal full-load torque is usually less than half the pull-out torque. If s_m denotes the value of slip corresponding to the maximum torque, then

$$\frac{T}{T_m} = \frac{2}{s/s_m + s_m/s}.$$ (6.30)

The value of the breakdown torque is determined by the total leakage reactance X, and the slip at which it occurs is determined by the rotor resistance R'_2. Fig. 6.8 shows a family of torque–speed curves for different values of R'_2,

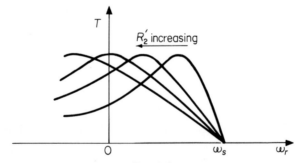

Fig. 6.8

and it will be seen that increasing the rotor resistance increases the standstill torque (i.e. the torque when the rotor is stationary). Unfortunately this also detracts from the full-load performance, for at small values of slip R'_2/s is very much greater than X, and equation 6.27 becomes

$$T = \frac{mpV_1^2}{\omega} \cdot \frac{s}{R'_2}.$$ (6.31)

Thus the torque–slip characteristic is linear in this region, with a slope inversely proportional to R_2'; and equation 6.31 shows that s must increase in proportion to R_2' if the machine is to develop the same full-load torque. The choice of R_2' therefore involves a compromise between the starting torque and the full-load slip, which in turn affects the efficiency.

CURRENT, POWER FACTOR AND EFFICIENCY

The form of the equivalent circuit (Fig. 6.4) shows that each phase of an induction motor will act as an inductive impedance, and the machine therefore takes current at a lagging power factor. Since the rotor impedance $R_2'/s + jx_2'$ varies in magnitude and phase angle with the slip s, the stator current I_1 will also vary with slip (and hence with the rotor speed ω_r). Figures 6.9b and 6.9c show the variation of the stator current magnitude I_1 and the power factor $\cos\phi$ (where ϕ is the phase angle between V_1 and I_1) for a typical small cage induction motor. The full-load speed is shown in the figure, and it will be seen that the starting current when the rotor is stationary is about 5 times as large as the full-load running current. Induction motors are frequently started simply by connecting the stator directly to the supply (direct-on-line or d.o.l. starting). When the resulting high starting current is unacceptable, the phase voltage is reduced either by the use of auto-transformers or series reactors, or by using star connection of the windings for starting and changing over to delta connection after the rotor has accelerated.

The efficiency of an induction motor is defined in the usual way as the ratio of the mechanical output power to the electrical input power. A typical efficiency curve is shown in Fig. 6.9d, and the flow of power through the machine may be traced in the same way as for the d.c. motor (section 2.2); for an m-phase machine we have:

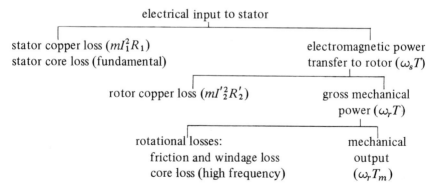

The important difference between this flow diagram and the corresponding diagram of the d.c. machine lies in the separation of the core loss into two

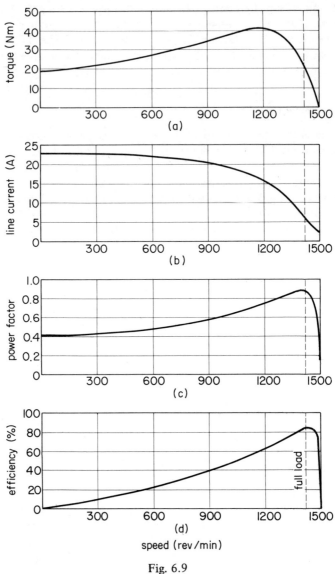

Fig. 6.9

components. The fundamental core loss, due to the fundamental component of the rotating magnetic field, is mainly confined to the stator core. This follows from the fact that the magnetic field at a point in the stator alternates with the supply angular frequency ω, whereas the corresponding field at a point fixed in the rotor alternates with the slip angular frequency $s\omega$; fundamental eddy-current and hysteresis losses are therefore insignificant in the rotor at the normal full-load slip. There is another component of core loss caused by (a) harmonic components of the rotating field which arise from the non-sinusoidal distribution of practical windings, and (b) pulsations in the field which arise from the relative motion of rotor and stator slots. This high-frequency component of core loss occurs in both the rotor and the stator, and the energy is supplied in a very complex way; it is customary to assume that it can be represented by a rotational loss term added to the mechanical (windage and friction) losses.

6.4 Speed control of induction motors

Induction motors account for about 90% of the electrical drives used in industry, and the majority of these applications require a fairly constant speed. Induction machines are normally designed to work with a small value of slip (generally less than 5%) at full load, and the deviation of the rotor speed ω_r from the synchronous speed ω_s is therefore small. There are certain applications, however, which require substantial variations in the motor speed. D.c. motors form an obvious choice for this kind of drive because of the ease of speed control; but they are relatively expensive, and the same objection applies to the variable-speed a.c. commutator motors*. The induction motor has the advantages of low cost and high reliability, and the possibility of controlling its speed is now examined.

The possible methods of speed control may be deduced from equation 6.19, which defines the fractional slip s. Thus

$$\omega_r = (1 - s)\omega_s = (1 - s)\frac{\omega}{p}, \qquad (6.32)$$

showing that the rotor speed may be controlled by varying the slip s, the number of pole-pairs p, or the supply angular frequency ω. These methods will be considered in turn.

VARIATION OF ROTOR SLIP

For a given load torque T, equation 6.28 shows that the quantity sX/R_2' is a constant. Increasing the resistance R_2' will cause a proportionate increase in the slip s, with a consequent decrease in the rotor speed. In practice the load torque will vary with the speed, and the precise effect of varying R_2' may be found by plotting the torque–speed characteristic for the load on the graph of motor torque–speed characteristics shown in Fig. 6.8. The point of intersection of the

* The best-known type is the Schrage motor, in which the speed is controlled by altering the position of the brushes. These machines are derived from the induction motor, but they are beyond the scope of this book.

load curve with the motor characteristic gives the speed for each value of R_2', as shown in Fig. 6.10. Because of the power loss associated with the slip, this is an inefficient method of speed control. It is often used for short periods when a large starting torque is required; a slipring motor is employed, and the external resistance is reduced to zero as the rotor runs up to speed. When continuous operation at high slip is required for speed control purposes, the slip power can

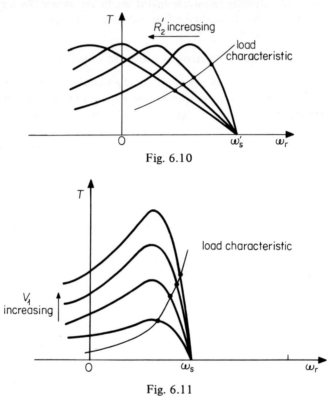

Fig. 6.10

Fig. 6.11

be extracted from the rotor circuit by connecting a second motor in place of the external resistance; this is the basis of the slip power recovery or Kramer system (Fitzgerald, Kingsley and Kusko [1]). An alternative method of varying the slip, which may be applied to cage rotor machines, is to vary the magnitude of the stator voltage V_1. Fig. 6.11 shows the family of torque-speed characteristics for a number of values of V_1, and the points of intersection with the load curve give the corresponding values of speed. As with rotor resistance variation, this is an inefficient method of speed control. It has the merit of simplicity, and it is sometimes used with small machines when efficiency is not particularly important.

POLE-CHANGE WINDINGS

   The second method of speed control is by alteration of the number of pole-pairs p; this can only give discrete changes of speed, since p must be an integer. With a properly designed cage rotor it is only necessary to alter the number of poles of the stator winding, for the corresponding rotor currents will find their own paths in the cage. An obvious way of varying p is to have an independent winding for each pole number, with a selector switch to connect the appropriate winding to the supply. A better solution is to design a single winding in such a way that the number of poles can be changed merely by altering the interconnection of the coils. The recently developed technique of pole–amplitude modulation (Jayawan [2]) permits values of p such as 4, 5, 6 to be obtained from a single stator winding, and this gives a useful degree of speed control.

FREQUENCY VARIATION

   The third and most interesting method of speed control is achieved by variation of the supply angular frequency ω. This permits continuous variation of the speed; the slip can be kept low to maintain the efficiency; and the method can be applied to cage induction motors. It is desirable to maintain a constant flux density in the air-gap when the frequency is varied, and equation 4.49 shows that the magnitude of the supply voltage must vary in proportion to the frequency if B_m is to remain constant. This may also be deduced from the equivalent circuit (Fig. 6.5), for the magnetizing current is given by

$$I_{0m} = \frac{V_1}{X_m} = \frac{V_1}{\omega L_m},$$
(6.33)

and this will be constant if $V_1 \propto \omega$. The torque is given by equation 6.28, and if

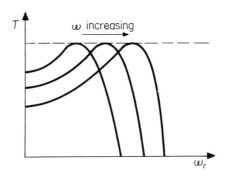

Fig. 6.12

the reactance X is written as ωL this becomes

$$T = \frac{mp}{L} \cdot \frac{V_1^2}{\omega^2} \cdot \frac{1}{s\omega L/R_2' + R_2'/s\omega L}. \tag{6.34}$$

Thus the breakdown torque is constant when $V_1 \propto \omega$, and a family of torque–speed characteristics for different frequencies is shown in Fig. 6.12. The variable-frequency stator supply may be obtained from the fixed-frequency a.c. mains via a solid-state frequency converter (Fitzgerald, Kingsley and Kusko [1], Hindmarsh [3]). The power semiconductor devices (transistors or thyristors) in the converter have to handle the full power of the induction motor, and this makes the system expensive at present; but the downward trend of semiconductor prices will soon make the variable-frequency induction motor a serious competitor to the controlled d.c. motor for industrial variable-speed drives.

6.5 Single-phase induction motors

Most industrial induction motors have three-phase stator windings, but for domestic applications it is necessary to operate induction motors from a single-phase supply (Alger [4]). One method is to use a two-phase machine, with a capacitor connected in series with one phase of the stator winding (Fig. 6.13). The current I_α in the phase connected directly to the supply will lag the supply voltage V by an angle α (Fig. 6.14). By suitable choice of the capacitance value,

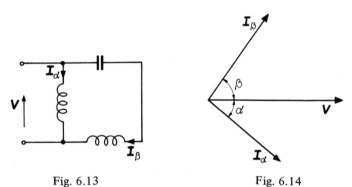

Fig. 6.13   Fig. 6.14

the current I_β in the other phase can be made to lead the voltage V by an angle β. If $\alpha + \beta = \pi/2$, the currents are in quadrature, and the motor will operate as a normal two-phase machine. Since the impedance presented by each phase of the stator winding varies with the load on the motor, the phase splitting will only be exact for one particular value of load torque, and there will be some unbalance between the phases at other load conditions. Machines of this kind are known as capacitor motors.

An alternative approach is to take a machine with a normal cage rotor, but only a single stator winding connected to the single-phase supply. If a current i flows in this winding, it will set up an m.m.f. given by equation 4.6:

$$F = ik \cos \theta.$$

Since i is a sinusoidal alternating quantity of the form

$$i = I_m \cos \omega t,$$

the m.m.f. produced by the winding will be

$$F = I_m k \cos \omega t \cos \theta. \tag{6.35}$$

By a trigonometric identity this may be written in the form

$$F = I_m k \frac{1}{2} \left\{ \cos(\omega t - \theta) + \cos(\omega t + \theta) \right\}. \tag{6.36}$$

$$\qquad\qquad \text{(a)} \qquad \text{(b)}$$

Term (a) in equation 6.36 represents a field rotating with angular velocity ω in the positive direction; while term (b) represents a field rotating with angular velocity ω in the negative direction. Thus the pulsating field produced by alternating current flowing in a single-phase winding may be resolved into two rotating fields which rotate in opposite directions. The machine behaves as though it had two polyphase windings, carrying the same current magnitude per phase, but producing magnetic fields which rotate in opposite directions. Each

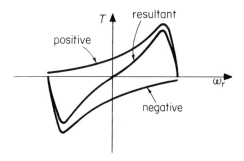

Fig. 6.15

rotating field will give rise to a torque on the rotor in the normal way, with a corresponding torque—speed characteristic. For the field rotating in the positive direction, this is the normal torque—speed curve; for the negative rotating field, the direction of torque is reversed, and zero slip now corresponds to a rotor speed of $-\omega_s$. Fig. 6.15 shows the two torque—speed curves, together with the resultant torque which is the sum of the two components. The resultant torque is zero when the rotor is stationary, so the machine is not inherently self-start-ing. If some means is provided for spinning the rotor, the resultant torque acts in a direction to accelerate the rotor, which runs up to speed in the normal way. The starting torque is provided by an auxiliary winding in space quadrature with the main winding, carrying a current which is approximately in time quadrature with the current in the main winding. The field produced by the auxiliary winding combines with a portion of the main field to produce a rotating field, which exerts a torque on the stationary rotor. There are three common arrangements for the auxiliary winding.

SHADED-POLE MOTOR

This simple form of single-phase motor is shown in Fig. 6.16. The stator has salient poles carrying the main winding, and a portion of each pole is enclosed by a ring which is usually made of copper. Currents are induced in the ring (or 'shading coil') by the alternating magnetic field, and the portion of the pole

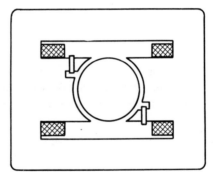

Fig. 6.16

enclosed by the ring is 'shaded' from the main pole flux; the flux is weaker, and its phase is retarded relative to the main flux. The arrangement forms a rudi-mentary two-phase winding which is adequate for accelerating the rotor against light loads. Motors of this kind are robust and inexpensive; but they are also inefficient, and their use is restricted to sizes below about 50 W.

SPLIT-PHASE MOTOR

For larger sizes of single-phase induction motor (up to about 1 kW) with modest starting torque requirements, a normal stator construction is used. The auxiliary winding is designed to have very different values of resistance and reactance from the main winding, so that there is an appreciable phase shift between the currents when the two windings are connected to the same single-phase supply. Thus the phase-splitting is inherent in the machine design, and the auxiliary winding normally has a short time rating. It is disconnected once the rotor has run up to speed, either by a centrifugally operated switch or by a relay which senses when the main winding current has fallen from a high starting value to the normal running value.

CAPACITOR-START MOTOR

When the motor must develop a large starting torque, the capacitor motor arrangement is used. This gives a larger phase angle between the main and auxiliary currents (ideally 90°), and permits a better design of the auxiliary winding since the phase shift is provided by the capacitor. As with the split-phase motor, the auxiliary winding is disconnected once the rotor has run up to speed.

Single-phase induction motors have lower values of efficiency and power factor than comparable polyphase machines, and their use is restricted to powers below about 4 kW. The majority have ratings below 1 kW, and they are employed in very large numbers. Most of the motors used in domestic appliances are single-phase induction motors, and some small industrial drives are also of this kind.

Problems

6.1 In problem 2.3, the d.c. shunt motor is replaced by a slipring induction motor with an external resistance R in each phase of the rotor circuit. If the internal losses of the motor may be neglected, show that the same expressions hold for the hoisting speed and the efficiency.

6.2 A cage induction motor has direct current applied to the stator winding. Show that the machine acts as a brake for both directions of rotation and deduce the shape of the torque−speed curve by considering the slip speed of an induction motor operating from an a.c. supply.

6.3 When the rotor of an induction motor accelerates from rest with no mechanical load coupled to the shaft, part of the energy input to the rotor will be dissipated as heat in the rotor resistance and part will be stored as kinetic energy of the rotor. If rotational losses may be neglected, show that when the rotor reaches the synchronous speed the total energy dissipated in the rotor resistance is equal to the final kinetic energy of the rotor.

6.4 An induction motor has a two-phase stator winding, and it runs with a slip s when positive-sequence voltages V_p and $-jV_p$ are applied to the α and β phases respectively. With the rotor speed unchanged the positive-sequence supply is disconnected and negative-sequence voltages V_n, jV_n are applied to the α and β phases respectively. Show that the slip is now equal to $2 - s$, and draw complete equivalent circuits for the two conditions of operation.

6.5 In problem 6.4, the voltages V_p and V_n are adjusted so that in the first case the α and β phase currents are $I, -jI$ and in the second they are I, jI. Use the principle of superposition to show that the motor operates as a single-phase machine when voltages $(V_p + V_n), (-jV_p + jV_n)$ are applied to the α and β phases respectively, and hence combine the two separate equivalent circuits into a single equivalent circuit for the single-phase machine. Note that the currents in a circuit will be doubled if all the impedance values are halved.

References

[1] FITZGERALD, A. E., KINGSLEY, C., Jr. and KUSKO, A. (1971). *Electric machinery*, 3rd edition. McGraw-Hill, New York.
[2] JAYAWANT, B. V. (1968). *Induction machines.* McGraw-Hill, Maidenhead.
[3] HINDMARSH, J. (1970). *Electrical machines and their applications,* 2nd edition. Pergamon, Oxford.
[4] ALGER, P. L. (1970). *Induction machines,* 2nd edition. Gordon & Breach, New York.

CHAPTER 7

Generalized Machine Theory

7.1 The unity of rotating machines

We have seen that there is a close relationship between the induction machine and the synchronous machine, and the synchronous machine was developed from the d.c. machine by omitting the commutator. We now show that all three may be regarded as particular cases of a more general type of machine known as a 'doubly fed' machine.

THE DOUBLY FED MACHINE

Consider a machine constructed in the same way as an induction machine, with polyphase windings on both the stator and the rotor. Let the stator be connected to a polyphase source having an angular frequency ω_1; if it is a two-pole winding, it will produce a magnetic field rotating with an angular velocity ω_1. Let the rotor be connected to a second polyphase source having an angular frequency ω_2; this will set up a magnetic field rotating with angular velocity ω_2 *with respect to the rotor*. If the rotor itself has an angular velocity ω_r, the speed *relative to the stator* of this rotating field will be $\omega_r + \omega_2$. The stator and rotor fields will lock together to produce a steady torque if, and only if

$$\omega_1 = \omega_r + \omega_2. \tag{7.1}$$

Note that ω_2 may be positive or negative, since the rotor field can be made to rotate in either direction by altering the phase sequence of the supply. The rotor speed is thus given by

$$\omega_r = |\omega_1| \pm |\omega_2|. \tag{7.2}$$

The doubly fed machine shares the properties of both synchronous and induction machines. It is synchronous in the sense that the rotor speed is determined absolutely by the stator and rotor frequencies; but the stator field can also induce voltages in the rotor winding, since the rotor speed differs from the speed of the stator rotating field. In the steady state there will be a constant angle δ_{12} between the stator and rotor magnetic axes, and from equation 4.23 the torque is given by

$$T = KB_{1m}B_{2m} \sin \delta_{12} \qquad (7.3)$$

where δ_{12} is the angle between the magnetic axes of the stator and rotor fields.

Power relationships

Power relationships may be deduced for this machine in the same way as for the induction machine. The electromagnetic power supplied by the stator winding is

$$P_1 = \omega_1 T, \qquad (7.4)$$

and the mechanical power output of the rotor is

$$P_r = \omega_r T. \qquad (7.5)$$

The electromagnetic power supplied by the rotor must therefore be

$$P_2 = P_r - P_1 = (\omega_r - \omega_1)T, \qquad (7.6)$$

and from equation 7.1 this is

$$P_2 = -\omega_2 T. \qquad (7.7)$$

Thus if ω_2 is positive, i.e. in the same direction as the stator field velocity ω_1, the rotor will run at the difference of the two speeds and there will be an electrical power *output* from the rotor winding. Only the difference between the electrical powers, $|P_2| - |P_1|$, will be converted into mechanical power. On the other hand, if ω_2 is negative, i.e. in the opposite direction to the stator field velocity ω_1, the rotor will run at the sum of the two speeds, and there will be an electrical power input to the rotor. An interesting possibility is to make $-\omega_2 = \omega_1 = \omega$ by supplying the rotor from the same source as the stator. Then

$$\omega_r = 2\omega, \qquad (7.8)$$

$$P_1 = P_2 = \tfrac{1}{2}P_r. \qquad (7.9)$$

The rotor runs at twice the normal synchronous speed, and the mechanical power output is supplied equally from the stator and the rotor; the machine would give twice the output of a conventional synchronous machine of the same size. Unfortunately such a machine exhibits a dynamic instability which results in loss of synchronism, and no satisfactory method has yet been found for stabilizing it (Bird, McCloy and Chalmers [1]).

PARTICULAR CASES OF THE DOUBLY FED MACHINE

We now show how the conventional machine types may be obtained from the general doubly fed machine by imposing certain constraints.

Induction machine

The rotor winding is connected to a resistive load; there is no longer any external constraint on the frequency ω_2, and this simply becomes the slip frequency, determined by the rotor speed ω_r. Removing the frequency constraint means that the rotor speed can take any value, and the actual operating speed is determined by the fact that the torque produced by the machine must exactly balance the mechanical load torque. The rotor power output $|P_2|$ is dissipated in the resistance, and for high efficiency this must be minimized by keeping the slip frequency ω_2 as low as possible. With the frequency ω_2 unconstrained, the angle δ_{12} is fixed by the electrical characteristics of the rotor circuit.

Synchronous machine

In the conventional synchronous machine, the rotor frequency ω_2 is zero since the excitation winding carries direct current. Thus the rotor speed is constrained to be equal to the stator angular frequency ω_1, and the rotor power P_2 is zero. (There will, of course, be a d.c. power loss in the rotor on account of its resistance; but there will be no transfer of electromagnetic power from the rotor winding.) With this constraint on the motor frequency, the load angle δ_{12} is not constrained; it is determined by the load connected to the machine shaft. The synchronous machine thus avoids the inherent inefficiency of the induction motor by eliminating the rotor loss $|P_2|$, and the d.c. loss can be avoided by using permanent magnets for the rotor field.

D.c. machine

A synchronous machine can be inverted by supplying d.c. to the stator and polyphase a.c. to the rotor of the doubly fed model. The machine then has a stationary field and a rotating armature; small machines are often made in this way. In this inverted form the synchronous machine bears a close resemblance to the d.c. machine, which also has a stationary field and a rotating armature. The d.c. machine can therefore be regarded as a synchronous machine together with a frequency changer in the form of a commutator, which converts the direct current entering the brushes into alternating current in the armature winding. The frequency of the rotor currents, ω_2, is now determined by the speed of the commutator – and hence by the speed of the rotor – and it is thus unconstrained. In place of a frequency constraint, the commutator constrains the angle δ_{12} to be $90°$, giving the maximum possible torque for a given stator and rotor field strength. Since B_{1m} is proportional to the field current and B_{2m} to the armature current, equation 7.3 reduces to the d.c. machine torque equation

$$T \propto i_f i_a. \tag{7.10}$$

GENERALIZED MACHINE THEORY

This essential unity of the different types of rotating machine suggests the possibility of a unified mathematical theory. Such a theory was developed by Kron in the 1930s, using the mathematical apparatus of the tensor calculus (Kron [2]). It is possible to express the basis of this theory in terms of matrices, and a key element is a mathematical transformation of variables which is

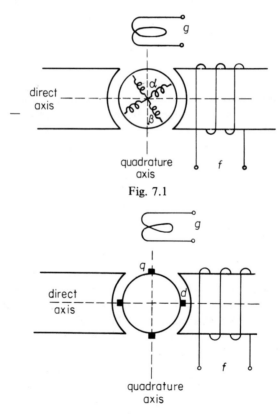

Fig. 7.1

Fig. 7.2

equivalent to the physical action of the commutator. This transformation enables the basic equations of a.c. and d.c. machines to be written in identical form, and a routine process permits these equations to be set up by inspection.

The model, or 'primitive', machine used in the generalized theory is an adaptation of the doubly fed machine already considered. Two forms of the primitive machine are used: a slipring model and a commutator model. The slipring model resembles our doubly fed machine in having polyphase windings on the stator and the rotor; for simplicity these are two-phase windings,* and for generality one member (the stator) can have salient poles (Fig. 7.1).† Connection to the rotor winding is made via sliprings. In the commutator model, the stator is the same, but the rotor carries a commutator winding with two sets of brushes (Fig. 7.2). We shall first set up the equations of this commutator, or d.c., machine model by a generalization of the equations already developed for simple d.c. machines. The equations of the slipring, or a.c., machine model will be established by treating the stator and rotor windings as coupled circuits with variable inductance coefficients, and we shall then unify the two theories by a transformation of variables.

7.2 D.c. machine equations

Fig. 7.3 is a symbolic representation of the commutator machine model shown in Fig. 7.2. It is a two-pole d.c. machine, with the normal field winding *f* wound on salient poles; the magnetic axis of these poles is known as the direct axis. The normal pair of brushes *q* lies on an axis known as the quadrature axis, at right angles to the direct axis. In addition, the model has a second field winding *g* on the quadrature axis, and a second pair of brushes *d* on the direct axis. This model represents the structure of the cross-field machines, with the ordinary d.c. machine as a special case.

The currents in the brushes and the field windings will set up fluxes Φ_d and Φ_q on the direct and quadrature axes. As in the normal d.c. machine, rotation of the armature will produce a generated voltage between the *q* brushes on account of Φ_d, and a generated voltage between the *d* brushes on account of

* There is no loss of generality in assuming a two-phase winding. It is shown in Appendix B that for any arbitrary three-phase currents flowing in a three-phase winding, we can always find the corresponding two-phase currents flowing in a two-phase winding which will produce exactly the same magnetic field in the machine air gap.

† A restriction of the generalized theory is that only one member (either the stator or the rotor) can have salient poles, and the windings on the non-salient member must be balanced (Hancock [3]). Nearly all practical machines meet these requirements; important exceptions are the inductor alternator (which has saliency on both sides of the air gap) and the single-phase alternator (which has an unbalanced armature winding).

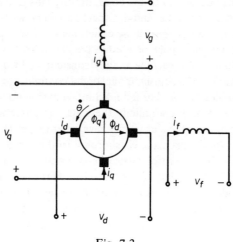

Fig. 7.3

Φ_q. If the fluxes Φ_d and Φ_q vary with time, there will be induced voltages in the d and q axis windings respectively, due to the rate of change of flux linkage; there will be no interaction between the axes, since they are at right angles.

VOLTAGE EQUATION FOR A FIELD WINDING

For the main field winding on the direct axis we have

$$v_f = R_f i_f + \frac{d\psi_f}{dt}$$

$$= R_f i_f + \frac{d}{dt}(L_f i_f + M_{df} i_d)$$

$$= (R_f + L_f p)i_f + M_{df} p i_d, \qquad (7.11)$$

where ψ_f is the flux linking the d-axis field winding (f), L_f is the self inductance of the winding, and M_{df} is its mutual inductance with the d-axis armature circuit. The Heaviside notation of p for the d/dt operator is used rather than D, to avoid confusion with d which denotes the direct axis.

VOLTAGE EQUATION FOR AN ARMATURE WINDING

For the brushes on the direct axis, we have

$$v_d = R_d i_d + \frac{d\psi_d}{dt} + K_a \Phi_q \dot{\theta}, \qquad (7.12)$$

where $\dot{\theta} = \omega$ is the angular velocity of the rotor, ψ_d is the flux linking the d-axis armature circuit and K_a is the fundamental armature constant. The flux Φ_q will depend on i_q and i_g, and with normal machine windings it may be shown (Jones [4]) that

$$K_a \Phi_q = L_q i_q + M_{qg} i_g, \qquad (7.13)$$

where L_q is the self-inductance of the q-axis armature circuit, and M_{qg} is the mutual inductance between this circuit and the q-axis field winding (g). Equation 7.12 therefore becomes

$$v_d = (R_d + L_d p)i_d + M_{df} p i_f + L_q \dot{\theta} i_q + M_{qg} \dot{\theta} i_g. \qquad (7.14)$$

MATRIX VOLTAGE EQUATION

Expressions similar to equations 7.11 and 7.14 may be written for the voltages v_g and v_q, and the set of four equations may then be put in matrix form:

$$
\begin{array}{c}
\\
d \\
q \\
f \\
g
\end{array}
\begin{bmatrix}
v_d \\
v_q \\
v_f \\
v_g
\end{bmatrix}
=
\begin{array}{cccc}
d & q & f & g
\end{array}
\begin{array}{c}
d \\
q \\
f \\
g
\end{array}
\begin{bmatrix}
R_d + L_d p & L_q \dot{\theta} & M_{df} p & M_{qg} \dot{\theta} \\
-L_d \dot{\theta} & R_q + L_q p & -M_{df} \dot{\theta} & M_{qg} p \\
M_{df} p & 0 & R_f + L_f p & 0 \\
0 & M_{qg} p & 0 & R_g + L_g p
\end{bmatrix}
\begin{array}{c}
d \\
q \\
f \\
g
\end{array}
\begin{bmatrix}
i_d \\
i_q \\
i_f \\
i_g
\end{bmatrix}
\quad (7.15)
$$

The labelling of the rows and columns was introduced by Kron, and provides a useful means of identifying the matrix elements (especially when numerical values are used in place of symbols). On account of the symmetry of the armature, $R_d = R_q = R_a$. But the magnetic circuits of the two axes are not in general identical, so that $L_d \neq L_q$. We may write the machine equations 7.15 in the form

$$v = Zi,$$

and the square matrix Z is known as the operational, or transient, impedance matrix of the machine.

7.3 A.c. machine equations

Consider the slipring model shown in Figs. 7.1. and 7.4. This has balanced two-phase windings α and β on the rotor, a main field winding f on the stator direct axis, and a second field winding g on the stator quadrature axis. With the

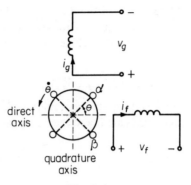

Fig. 7.4

winding g omitted this represents the normal synchronous machine; with no saliency (i.e. a uniform air gap) and identical f and g windings, the model represents the two-phase induction machine. We may write the voltage equation for the α-phase winding as

$$v_\alpha = R_\alpha i_\alpha + \frac{d\psi\alpha}{dt}$$

$$= R_\alpha i_\alpha + p(L_\alpha i_\alpha + M_{\alpha\beta} i_\beta + M_{\alpha f} i_f + M_{\alpha g} i_g). \tag{7.16}$$

VARIABLE INDUCTANCE COEFFICIENTS

In general, $L_\alpha, M_{\alpha\beta}, M_{\alpha f}$ and $M_{\alpha g}$ are periodic functions of θ; to a first approximation we may take the first terms in the Fourier expansions of these functions. Thus L_α, the self inductance of the winding, will vary from a maximum value L_d when the winding is aligned with the direct axis ($\theta = 0$ or π), to a minimum value L_q when the winding is aligned with the quadrature axis. The simplest sine or cosine function to meet this requirement is

$$L_\alpha = \tfrac{1}{2}(L_d + L_q) + \tfrac{1}{2}(L_d - L_q) \cos 2\theta, \tag{7.17}$$

and this is equivalent to the assumption of a sinusoidally distributed winding: a good approximation which we have already used. The mutual inductance $M_{\alpha\beta}$ likewise is periodic in 2θ, and it may be shown that

$$M_{\alpha\beta} = \tfrac{1}{2}(L_d - L_q) \sin 2\theta. \tag{7.18}$$

The mutual inductances $M_{\alpha f}$ and $M_{\alpha g}$ are clearly periodic in θ, and we may put

$$M_{\alpha f} = M_{df} \cos \theta, \tag{7.19}$$

$$M_{\alpha g} = M_{qg} \sin \theta. \tag{7.20}$$

In these equations, M_{df} is the maximum value of $M_{\alpha f}$ when the α phase is aligned with the d axis, and M_{qg} is the maximum value of $M_{\alpha g}$ when the α phase is aligned with the q axis. Equations 7.17 and 7.18 may be re-written as

$$L_\alpha = L_d \cos^2 \theta + L_q \sin^2 \theta, \tag{7.21}$$

$$M_{\alpha\beta} = (L_d - L_q) \sin \theta \cos \theta, \tag{7.22}$$

and the voltage equation 7.16 becomes

$$v_\alpha = R_\alpha i_\alpha + p(L_d \cos^2 \theta + L_q \sin^2 \theta) i_\alpha + (L_d - L_q) p \sin \theta \cos \theta \, i_\beta +$$
$$+ M_{df} p \cos \theta \, i_f + M_{qg} p \sin \theta \, i_g. \tag{7.23}$$

MATRIX VOLTAGE EQUATION

The voltage equations for the other three windings can be derived in the same way, giving the following matrix equation:

$$
\begin{bmatrix} v_\alpha \\ v_\beta \\ v_f \\ v_g \end{bmatrix}
=
\begin{array}{c} \alpha \\ \beta \\ f \\ g \end{array}
\begin{bmatrix}
\begin{array}{c} R_\alpha + \\ p(L_d \cos^2 \theta + L_q \sin^2 \theta) \end{array} & (L_d - L_q)p \sin \theta \cos \theta & M_{df} p \cos \theta & M_{qg} p \sin \theta \\
(L_d - L_q)p \sin \theta \cos \theta & \begin{array}{c} R_\beta + \\ p(L_d \sin^2 \theta + L_q \cos^2 \theta) \end{array} & M_{df} p \sin \theta & M_{qg} p \cos \theta \\
M_{df} p \cos \theta & M_{df} p \sin \theta & R_f + L_f p & \\
M_{qg} p \sin \theta & M_{qg} p \cos \theta & & R_g + L_g p
\end{bmatrix}
\begin{bmatrix} i_\alpha \\ i_\beta \\ i_f \\ i_g \end{bmatrix}
$$

$$\tag{7.24}$$

In these equations, $R_\alpha = R_\beta = R_a$ since the armature winding is balanced.

COMMUTATOR TRANSFORMATION

To simplify the equations we introduce a change of variables which will remove the circular functions of θ from the impedance matrix. There are many possible transformations which will achieve this result; we select a transformation such that, voltage and current transform in the same way, and the total instantaneous power (Σvi) is unchanged by the transformation. It may be shown (Hancock [3]) that a transformation to meet these requirements is given by

$$
\begin{matrix} & \alpha & \beta \end{matrix}
$$

$$
\begin{matrix} d \\ q \end{matrix}\begin{bmatrix} v_d \\ v_q \end{bmatrix} = \begin{matrix} d \\ q \end{matrix}\begin{bmatrix} \cos\theta & \sin\theta \\ \sin\theta & -\cos\theta \end{bmatrix}\begin{matrix} \alpha \\ \beta \end{matrix}\begin{bmatrix} v_\alpha \\ v_\beta \end{bmatrix}, \tag{7.25}
$$

$$
\begin{matrix} & \alpha & \beta \end{matrix}
$$

$$
\begin{matrix} d \\ q \end{matrix}\begin{bmatrix} i_d \\ i_q \end{bmatrix} = \begin{matrix} d \\ q \end{matrix}\begin{bmatrix} \cos\theta & \sin\theta \\ \sin\theta & -\cos\theta \end{bmatrix}\begin{matrix} \alpha \\ \beta \end{matrix}\begin{bmatrix} i_\alpha \\ i_\beta \end{bmatrix}. \tag{7.26}
$$

The square matrix in equations 7.25 and 7.26 is termed the commutator transformation matrix, for a reason which will be explained shortly. If we denote this
matrix by C, it has the useful property that $C = C^T = C^{-1}$; thus the same transformation gives the α, β quantities in terms of the d, q quantities. On making the
substitutions indicated by equations 7.25 and 7.26, the matrix equation 7.24
becomes

$$
\begin{matrix} & d & q & f & g & \\ d \\ q \\ f \\ g \end{matrix}\begin{bmatrix} v_d \\ v_q \\ v_f \\ v_g \end{bmatrix} = \begin{matrix} d \\ q \\ f \\ g \end{matrix}\begin{bmatrix} R_a + L_d p & L_q\dot\theta & M_{df}p & M_{qg}\dot\theta \\ -L_d\dot\theta & R_a + L_q p & -M_{df}\dot\theta & M_{qg}p \\ M_{df}p & 0 & R_f + L_f p & 0 \\ 0 & M_{qg}p & 0 & R_g + L_g p \end{bmatrix}\begin{matrix} d \\ q \\ f \\ g \end{matrix}\begin{bmatrix} i_d \\ i_q \\ i_f \\ i_g \end{bmatrix} \tag{7.27}
$$

This is identical to the matrix equation 7.15 for the d.c. machine. Thus the
transformation to d–q variables is physically equivalent to replacing the two-
phase armature winding by a commutator winding with brushes on the d and q
axes. The currents i_d and i_q are the currents which would have to flow in the
brushes to give the same machine performance. An a.c. machine represented in
d–q variables is shown symbolically in Fig. 7.5.

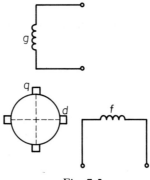

Fig. 7.5

7.4 General equations

GENERAL VOLTAGE EQUATION

We have seen that the equations for the d.c. machine and for the a.c. machine in terms of $d-q$ variables have the same form, and may be written as

$$v = Zi. \qquad (7.28)$$

Three different kinds of quantity occur in the elements of the impedance matrix Z: constants, which denote the winding resistances; self or mutual inductance terms, of the form Lp; and generated voltage terms, of the form $L\theta$. The impedance matrix may thus be expressed in terms of three matrices:

$$Z = R + Lp + G\theta, \qquad (7.29)$$

where R, L and G all have constant elements, and are given by

$$R = \begin{array}{c} \\ d \\ q \\ f \\ g \end{array} \overset{\begin{array}{cccc} d & q & f & g \end{array}}{\begin{bmatrix} R_a & & & \\ & R_a & & \\ & & R_f & \\ & & & R_g \end{bmatrix}}$$

$$L = \begin{array}{c} \\ d \\ q \\ f \\ g \end{array} \overset{\begin{array}{cccc} d & q & f & g \end{array}}{\begin{bmatrix} L_d & & M_{df} & \\ & L_q & & M_{qg} \\ M_{df} & & L_f & \\ & M_{qg} & & L_g \end{bmatrix}}$$

$$G = \begin{array}{c} \\ d \\ q \\ f \\ g \end{array} \overset{\begin{array}{cccc} d & q & f & g \end{array}}{\begin{bmatrix} & L_q & & M_{qg} \\ -L_d & & -M_{df} & \\ & & & \\ & & & \end{bmatrix}}$$

The resolution of Z into three parts reveals a simple structure, and Hancock [3] gives the following rules for writing down by *inspection* the impedance matrix for a machine with any number of windings:

(1)   Write in the principal diagonal the terms representing the resistances of the windings.

(2)   Also along the principal diagonal write the Lp terms corresponding to the self inductances of the windings.

(3)   Write in the appropriate places the mutual inductance terms Mp wherever the windings have a common axis. Each term will appear twice since mutual terms are always symmetric.

(4)   Wherever Lp or Mp occurs in a d or q row, write $L\dot\theta$ or $M\dot\theta$ in the same column of the other q or d row, prefixing those in the q row with a negative sign.

POWER IN TERMS OF MATRICES

For the d.c. machine in section 7.1, the total instantaneous power input to the windings is

$$P = i_d v_d + i_q v_q + i_f v_f + i_g v_g. \tag{7.30}$$

This may be expressed as a product of the voltage matrix with the transposed current matrix:

$$P = \begin{bmatrix} i_d & i_q & i_f & i_g \end{bmatrix} \begin{bmatrix} v_d \\ v_q \\ v_f \\ v_g \end{bmatrix} \tag{7.31}$$

$$= i^T v.$$

For the a.c. machine of section 7.2, the instantaneous input power is likewise $P = i^T v$, where the elements of i and v are the actual winding quantities. The commutator transformation was chosen to satisfy the condition

$$i_\alpha v_\alpha + i_\beta v_\beta = i_d v_d + i_q v_q, \tag{7.32}$$

so the power is also $P = i^T v$ in terms of the $d-q$ variables.

TORQUE EQUATION

In terms of the components of Z, equation 7.27 becomes

$$v = R\,i + Lpi + G\dot\theta i. \tag{7.33}$$

Pre-multiply this equation by i^T:

$$i^T v = i^T R\,i + i^T Lpi + i^T G\dot\theta i. \tag{7.34}$$

In equation 7.34, the terms have the following significance:

$i^T v$ is the total instantaneous input power.

$i^T R i$ is of the form $\Sigma R i^2$ since R is diagonal; it therefore represents the total copper loss in the machine.

$i^T L p i$ is the sum of terms of the form $iLpi = p(\frac{1}{2}Li^2)$ and $i_1 M p i_2 + i_2 M p i_1 = p(M i_1 i_2)$; it therefore represents the rate of change of stored magnetic energy.

$i^T G \dot{\theta} i$ is the difference between the input power and the other two terms; it represents the mechanical output power.

The following can therefore be written

$$\text{mechanical output} = T\dot{\theta} = i^T G \dot{\theta} i,$$

$$\text{i.e. } T = i^T G i. \tag{7.35}$$

This is one form of the general torque equation for electrical machines. As a particular case, take the simple d.c. machine. The impedance matrix is obtained by deleting the g and d rows and columns from equations 7.15.

$$Z = \begin{array}{c} q \\ f \end{array}\begin{array}{c} q \qquad\qquad f \\ \begin{bmatrix} R_a + L_q p & -M_{df}\dot{\theta} \\ & R_f + L_f p \end{bmatrix} \end{array}. \tag{7.36}$$

Thus

$$G = \begin{array}{c} q \\ f \end{array}\begin{array}{c} q \qquad f \\ \begin{bmatrix} & -M_{df} \\ & \end{bmatrix} \end{array}, \tag{7.37}$$

and

$$T = i^T G i = -M_{df} i_f i_q, \tag{7.38}$$

which is equivalent to the equation

$$T = K i_f i_a \tag{2.14}$$

developed earlier for the simple d.c. machine.

7.5 Performance calculations

The equations of a d.c. machine with any number of field windings are readily obtained from the general matrix equations, and it is a straightforward matter to solve any particular problem involving interconnection of the windings by a device described in section 7.6. A.c. machines are not so straightforward, and the method is best treated by way of an example: the steady-state operation of a salient-pole synchronous machine.

SALIENT-POLE SYNCHRONOUS MACHINE

Consider a machine with only one field winding, placed on the d axis (Fig. 7.6). The matrix voltage equation is obtained from equation 7.27 by

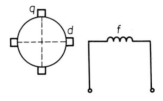

Fig. 7.6

omitting the g row and column; this eliminates the mutual inductance M_{qg} and the remaining mutual inductance M_{df} may be written as M:

$$
\begin{array}{c} d \\ q \\ f \end{array}
\begin{bmatrix} v_d \\ v_q \\ v_f \end{bmatrix}
=
\begin{array}{c} d \\ q \\ f \end{array}
\overset{\displaystyle \begin{array}{ccc} d & q & f \end{array}}{
\begin{bmatrix}
R_a + L_d p & L_q \dot{\theta} & Mp \\
-L_d \dot{\theta} & R_a + L_q p & -M\dot{\theta} \\
Mp & & R_f + L_f p
\end{bmatrix}}
\begin{array}{c} d \\ q \\ f \end{array}
\begin{bmatrix} i_d \\ i_q \\ i_f \end{bmatrix}
\qquad (7.39)
$$

For balanced steady state operation the phase voltages and currents will be sinusoidal quantities, and we may write:

$$v_\alpha = \sqrt{2}V \sin \omega t, \qquad (7.40)$$

$$v_\beta = \sqrt{2}V \sin(\omega t - \pi/2) = -\sqrt{2}V \cos \omega t, \qquad (7.41)$$

$$i_\alpha = \sqrt{2}I \sin(\omega t - \phi), \qquad (7.42)$$

$$i_\beta = \sqrt{2}I \sin(\omega t - \phi - \pi/2) = -\sqrt{2}I \cos(\omega t - \phi). \quad (7.43)$$

where V and I are the r.m.s. values of the armature voltage and current. The armature rotates with a steady angular velocity ω; since $\dot{\theta} = \omega$, we have

$$\theta = \omega t - \delta \qquad (7.44)$$

where δ is the angle between the axis of the α phase and the d axis at time $t = 0$.

Two-axis voltage equations

The commutator transformation (equation 7.25) gives

$$\begin{matrix} & \alpha & \beta \end{matrix}$$

$$d\begin{bmatrix} v_d \\ v_q \end{bmatrix} = d\begin{bmatrix} \cos(\omega t - \delta) & \sin(\omega t - \delta) \\ \sin(\omega t - \delta) & -\cos(\omega t - \delta) \end{bmatrix} \begin{matrix} \alpha \\ \beta \end{matrix}\begin{bmatrix} \sqrt{2}V \sin \omega t \\ -\sqrt{2}V \cos \omega t \end{bmatrix} = d\begin{bmatrix} \sqrt{2}V \sin \delta \\ \sqrt{2}V \cos \delta \end{bmatrix} \tag{7.45}$$

Similarly
$$i_d = -\sqrt{2}I \sin(\phi - \delta), \tag{7.46}$$

$$i_q = \sqrt{2}I \cos(\phi - \delta). \tag{7.47}$$

Thus v_d, v_q, i_d and i_q are all steady (d.c.) quantities. For steady-state operation v_f and i_f will both be steady d.c. quantities; all the time derivatives in equation 7.39 will vanish, and the equations may be written in the form

$$\begin{matrix} & d & g & f \end{matrix}$$

$$\begin{matrix} d \\ q \\ f \end{matrix}\begin{bmatrix} \sqrt{2}V \sin \delta \\ \sqrt{2}V \cos \delta \\ V_f \end{bmatrix} = \begin{matrix} d \\ q \\ f \end{matrix}\begin{bmatrix} R_a & X_q & \\ -X_d & R_a & -X_m \\ & & R_f \end{bmatrix}\begin{matrix} d \\ q \\ f \end{matrix}\begin{bmatrix} -\sqrt{2}I \sin(\phi - \delta) \\ \sqrt{2}I \cos(\phi - \delta) \\ I_f \end{bmatrix} \tag{7.48}$$

where $X_d = \dot{\theta}L_d = \omega L_d$, etc. Thus

$$V \sin \delta = -R_a I \sin(\phi - \delta) + X_q I \cos(\phi - \delta), \tag{7.49}$$

$$V \cos \delta = X_d I \sin(\phi - \delta) + R_a I \cos(\phi - \delta) - X_m I_f/\sqrt{2} \tag{7.50}$$

With the armature on open circuit, $I = 0$; and from equation 7.49, $\delta = 0$. If we let $V = E$ under these conditions, then equation 7.50 gives

$$E = [V]_{I=0} = -X_m I_f/\sqrt{2}. \tag{7.51}$$

Equations 7.49 and 7.50 now become

$$V \sin \delta = X_q I \cos(\phi - \delta) - R_a I \sin(\phi - \delta), \tag{7.52}$$

$$V \cos \delta = E + X_d I \sin(\phi - \delta) + R_a I \cos(\phi - \delta). \tag{7.53}$$

These are the basic equations for the steady-state operation of a synchronous machine. If we neglect armature resistance by putting $R_a = 0$, and remove saliency by letting $X_d = X_q = X_s$, then equations 7.52 and 7.53 reduce to

$$V \sin \delta = X_s I \cos(\phi - \delta), \tag{7.54}$$

$$V \cos \delta = E + X_s I \sin(\phi - \delta). \tag{7.55}$$

These equations may be obtained from the phasor diagram for the synchronous machine derived earlier (Fig. 5.8).

Torque equation

The torque is given by equation 7.35; from equation 7.39 the matrix G is

$$
G = \begin{matrix} & d & q & f \\ d \\ q \\ f \end{matrix}\begin{bmatrix} & L_q & \\ -L_d & & -M \\ & & \end{bmatrix}, \tag{7.56}
$$

and the torque is therefore

$$T = i^T G\, i = L_q i_q i_d - L_d i_d i_q - M i_f i_q \tag{7.57}$$

$$= \frac{1}{\omega}\left\{ i_d i_q (X_d - X_q) - i_f i_q X_m \right\}.$$

If R_a can be neglected, equations 7.52 and 7.53 become

$$V \sin \delta = X_q I \cos(\phi - \delta), \tag{7.58}$$

$$V \cos \delta = E + X_d I \sin(\phi - \delta). \tag{7.59}$$

From equations 7.46 and 7.47, these equations may be written as

$$\sqrt{2}V \sin \delta = X_q i_q, \tag{7.60}$$

$$\sqrt{2}V \cos \delta = E - X_d i_d, \tag{7.61}$$

and

$$\sqrt{2}E = -X_m i_f. \tag{7.51}$$

Substitution for i_d, i_q and i_f from these equations into equation 7.57 gives the torque:

$$T = \frac{1}{\omega}\left[2\,\frac{VE}{X_d}\sin \delta + V^2\left\{\frac{1}{X_q} - \frac{1}{X_d}\right\}\sin 2\delta \right]. \tag{7.62}$$

If $X_d = X_q$, this equation reduces to the torque expression derived earlier for the cylindrical rotor synchronous machine (equation 5.8), with $m = 2$ and $p = 1$; this is the synchronous torque

$$T_{syn} = \frac{2}{\omega X_d} VE \sin \delta. \tag{7.63}$$

When the rotor has salient poles so that $X_d \neq X_q$, there is an additional component of torque given by

$$T_{rel} = \frac{V^2}{\omega}\left\{\frac{1}{X_q} - \frac{1}{X_d}\right\}\sin 2\delta. \tag{7.64}$$

This is known as the reluctance torque; it is present in an unexcited synchronous machine (where E and hence T_{syn} are zero); and its magnitude is proportional to the difference between the reluctances of the rotor d-axis and q-axis magnetic circuits, since the reactance of a winding is inversely proportional to the reluctance of its magnetic circuit.

7.6 The connection matrix

So far we have found the equations for machines in which there are no connections between the windings. It often happens that some of the windings are interconnected, e.g. the shunt and series d.c. motors. A more complex example is the amplidyne (Fig. 7.7), and the aim of the analysis is to set up the equations

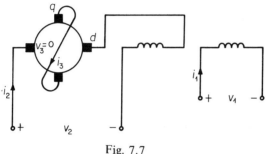

Fig. 7.7

relating the voltages and currents for machines with any arbitrary interconnection of the windings. Thus for the amplidyne, we want to find a matrix equation of the form

$$v' = Z' i', \tag{7.65}$$

where
$$v' = \begin{array}{c} 1 \\ 2 \\ 3 \end{array}\begin{bmatrix} v_1 \\ v_2 \\ 0 \end{bmatrix} \qquad i' = \begin{array}{c} 1 \\ 2 \\ 3 \end{array}\begin{bmatrix} i_1 \\ i_2 \\ i_3 \end{bmatrix}, \tag{7.66}$$

and Z' is as yet undetermined. Kron [5] has introduced a powerful method for handling this kind of problem, which may be described as follows. Let us remove the interconnections, and apply voltages to all the windings such that the currents remain unchanged (Fig. 7.8). The performance of the machine will be unchanged; the performance equations for the unconnected machine can be written down in the form

$$v = Z i, \tag{7.67}$$

and by expressing the individual winding currents i in terms of the system

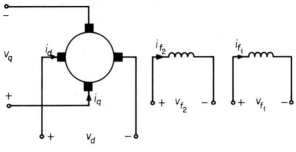

Fig. 7.8

currents i' it is possible to obtain Z' from Z. The procedure is as follows. We have:

$$i_d = i_2$$

$$i_q = i_3$$

$$i_{f_1} = i_1$$

$$i_{f_2} = -i_2$$

These relations may be written in matrix form:

$$
\begin{matrix}
 & & & 1 & 2 & 3 & \\
d\begin{bmatrix} i_d \\ i_q \\ i_{f_1} \\ i_{f_2} \end{bmatrix} & = & \begin{matrix} d \\ q \\ f_1 \\ f_2 \end{matrix}\begin{bmatrix} & 1 & \\ & & 1 \\ 1 & & \\ & -1 & \end{bmatrix} & \begin{matrix} 1 \\ 2 \\ 3 \end{matrix}\begin{bmatrix} i_1 \\ i_2 \\ i_3 \end{bmatrix}
\end{matrix}
$$

or $$i = C i',$$ (7.68)

where C is termed the connection matrix.

TRANSFORMATION OF THE IMPEDANCE MATRIX

Since the winding currents are unchanged, the instantaneous input power must be the same for the connected machine as for the unconnected machine. Thus,

$$(i')^T v' = i^T v.$$ (7.69)

Now $$i = C i',$$

so $\qquad i^T = (C\,i')^T = (i')^T C^T.$ (7.70)

Thus $\qquad (i')^T v' = i^T v = (i')^T C^T v.$ (7.71)

If equation 7.71 is to hold for any arbitrary $(i')^T$, then

$$v' = C^T v.$$ (7.72)

Now $\qquad v = Z\,i = Z\,C\,i'.$

Thus $\qquad Z'\,i' = v' = C^T v = C^T Z\,C\,i',$ (7.73)

and if equation 7.73 is to hold for any arbitrary i', we have

$$Z' = C^T Z\,C.$$ (7.74)

This is the transformation for obtaining the required impedance matrix Z' for the connected system from the known impedance matrix Z of the unconnected machine. Thus for the amplidyne we have

$$
Z =
\begin{array}{c}
 \\
d \\
q \\
f_1 \\
f_2
\end{array}
\begin{array}{cccc}
d & q & f_1 & f_2 \\
\left[\begin{array}{cccc}
R_a + L_d p & L_q \dot\theta & M_{df_1} p & M_{df_2} p \\
-L_d \dot\theta & R_a + L_q p & -M_{df_1}\dot\theta & -M_{df_2}\dot\theta \\
M_{df_1} p & & R_{f_1} + L_{f_1} p & M_{f_1 f_2} p \\
M_{df_2} p & & M_{f_1 f_2} p & R_{f_2} + L_{f_2} p
\end{array}\right]
\end{array}
$$ (7.75)

and carrying out the multiplication indicated in equation 7.74 gives the required matrix:

$$
Z' =
\begin{array}{c}
 \\
1 \\
2 \\
3
\end{array}
\begin{array}{ccc}
1 & 2 & 3 \\
\left[\begin{array}{ccc}
R_{f_1} + L_{f_1} p & (M_{df_1} - M_{f_1 f_2})p & \\
(M_{df_1} - M_{f_1 f_2})p & \begin{array}{c}R_a + R_{f_2} + \\ (L_d + L_{f_2} - 2M_{df_2})p\end{array} & L_q \dot\theta \\
-M_{df_1}\dot\theta & (M_{df_2} - L_d)\dot\theta & R_a + L_q p
\end{array}\right]
\end{array}.
$$ (7.76)

INTERCONNECTION OF MACHINES

The interconnection of two d.c. machines can be handled in a similar way. Let the equations of the unconnected machines be

$$v_1 = Z_1 i_1, \qquad v_2 = Z_2 i_2.$$ (7.77)

These may be written in a single compound matrix equation:

$$\begin{bmatrix} v_1 \\ v_2 \end{bmatrix} = \begin{bmatrix} Z_1 & \\ & Z_2 \end{bmatrix} \begin{bmatrix} i_1 \\ i_2 \end{bmatrix}, \qquad (7.78)$$

or
$$v = Z\,i. \qquad (7.79)$$

The currents in the connected system may be expressed as the elements of a matrix i', and the voltages as elements of a matrix v'. Equating winding currents in the connected and unconnected systems gives

$$i = C\,i', \qquad (7.80)$$

and the impedance matrix Z' for the system is thus

$$Z' = C^T Z\,C. \qquad (7.81)$$

The interconnection of any number of d.c. machines can be handled in the same way, by combining the individual voltage, current and impedance matrices into single compound matrices, related by equation 7.79, and applying equations 7.80 and 7.81.

The interconnection of a.c. machines is inherently more complex, since the physical interconnections between the windings set up relations between the actual phase currents and voltages (the α–β quantities), whereas the equations are expressed in terms of the transformed d–q quantities. It is not, in general, possible to equate the d–q quantities to obtain a connection matrix; the interconnection has to be handled in α–β terms and then transformed to d–q form. For two synchronous machines the result is a connection matrix C containing a variable $\theta = \theta_2 - \theta_1$, where θ_1 and θ_2 are the rotor angles of the two machines. The impedance transformation (equation 7.72) no longer holds when the elements of C are not constant; the correct expression is

$$Z' = C^T Z\,C + C^T L\,\frac{\partial C}{\partial \theta}\,\dot{\theta}. \qquad (7.82)$$

The theory is then conveniently handled with tensors (Lynn [6]); though for steady-state operation θ is constant; the second term on the right in equation 7.82 vanishes, and the equation reduces to the form of equation 7.81.

7.7 Conclusion

The generalized theory outlined in this chapter is a powerful analytical aid in the more complex machine problems, such as the transient and unbalanced operation of a.c. machines. Equally important is the unification of electrical machine theory; the same equations and methods apply to all machines which satisfy the basic assumptions of the theory, and this includes nearly all machines of practical importance. It is true that the general theory is a rather cumbersome tool for simple problems such as the d.c. shunt and series motors, but the 'break-even' point is quickly reached; the analysis of the salient-pole synchronous machine given in this chapter compares favourably with the classical steady-state theory, and it has the great advantage of being readily extended to cover transient conditions.

Two features of the generalized theory should be emphasized. One is the generality of the basic equations; they apply to all conditions of operation, with the steady state as a particular case. The second feature is the systematic way in which the equations are formulated. The impedance matrix for a machine of any complexity can be written down by inspection, with the known structure of the matrix forming an unequivocal check of its correctness. Interconnection of the windings (or of several machines) is then handled with matrix algebra, in which the operations are simple routine processes that are easily checked. It is thus possible to arrive at the final equations which specify the performance of a new machine or machine system with the certainty that these equations are correct. The solution of the equations is, of course, another matter!

Problems

7.1 (a) The general equations of the a.c. machine can be reduced to the two-phase induction motor by postulating a smooth (rather than salient) field structure and balanced field windings, one on each axis. If the rotor (armature) windings are short-circuited, show that the matrix equation can be written in the form

$$
\begin{array}{c}
\begin{array}{cc} & \end{array} \\
\begin{array}{c} d \\ q \\ f \\ g \end{array}
\begin{bmatrix} 0 \\ 0 \\ v_f \\ v_g \end{bmatrix}
=
\begin{array}{c} d \\ q \\ f \\ g \end{array}
\begin{array}{cccc} d & q & f & g \end{array}
\begin{bmatrix}
R_a + L_a p & L_a \dot{\theta} & Mp & M\dot{\theta} \\
-L_a \dot{\theta} & R_a + L_a p & -M\dot{\theta} & Mp \\
Mp & & R_b + L_b p & \\
& Mp & & R_b + L_b p
\end{bmatrix}
\begin{array}{c} d \\ q \\ f \\ g \end{array}
\begin{bmatrix} i_d \\ i_q \\ i_f \\ i_g \end{bmatrix}
\end{array}
$$

(b) Let a sinusoidal two-phase voltage source be connected to the f and g windings, so that

$$v_f = V_m \cos(\omega t + \phi) = Re\ Ve^{j\omega t}$$

$$v_g = V_m \cos(\omega t + \phi - \pi/2) = Re\ -jVe^{j\omega t}.$$

Under steady-state conditions, with the rotor running at a steady speed $\dot{\theta} = \omega_r$, the currents will all be sinusoidal alternating quantities. Put $i_d = \mathrm{Re}I_d\, e^{j\omega t}$ etc, and show that the matrix voltage equation becomes

$$
\begin{array}{c}
\\
d \\ q \\ f \\ g
\end{array}
\begin{bmatrix}
0 \\ 0 \\ V \\ -jV
\end{bmatrix}
=
\begin{array}{c}
\\
d \\ q \\ f \\ g
\end{array}
\begin{bmatrix}
\overset{d}{R_a + jX_a} & \overset{q}{(1-s)X_a} & \overset{f}{jX_m} & \overset{g}{(1-s)X_m} \\
(s-1)X_a & R_a + jX_a & (s-1)X_m & jX_m \\
jX_m & & R_b + jX_b & \\
& jX_m & & R_b + jX_b
\end{bmatrix}
\begin{array}{c}
\\
d \\ q \\ f \\ g
\end{array}
\begin{bmatrix}
I_d \\ I_q \\ I_f \\ I_g
\end{bmatrix}
$$

where $X_a = \omega L_a$, $X_b = \omega L_b$, $X_m = \omega M$, $s = (\omega - \omega_r)/\omega$.

(c) Show that the equations of part (b) are satisfied by

$$I_d = I_a, \qquad I_q = -jI_a$$
$$I_f = I_b, \qquad I_g = -jI_b$$

and hence that the machine equations are

$$0 = (R_a/s + jX_a)I_a + jX_mI_b,$$
$$V = jX_mI_a + (R_b + jX_b)I_b.$$

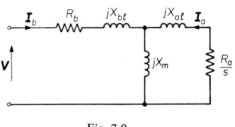

Fig. 7.9

(d) With the machine equations derived in part (c), let

$$X_a = X_{al} + X_m,$$
$$X_b = X_{bl} + X_m$$

Show that the equations now represent the circuit equations of Fig. 7.9, and interpret this result.

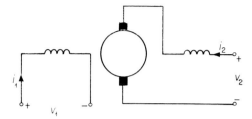

Fig. 7.10

7.2 Fig. 7.10 shows a d.c. compound motor. Show that the steady-state impedance matrix of the unconnected machine is

$$
\begin{array}{c}
\quad\quad q \quad\quad f_1 \quad\quad f_2 \\
\begin{array}{c} q \\ f_1 \\ f_2 \end{array}
\left[
\begin{array}{ccc}
R_a & -K_1\dot\theta & -K_2\dot\theta \\
 & R_{f_1} & \\
 & & R_{f_2}
\end{array}
\right]
\end{array}
$$

Hence show that the steady-state voltage equation for the connected machine is

$$
\begin{array}{c}
\quad\quad\quad\quad 1 \quad\quad\quad\quad\quad 2 \\
\begin{array}{c} 1 \\ 2 \end{array}
\left[\begin{array}{c} v_1 \\ v_2 \end{array}\right]
\begin{array}{c} 1 \\ 2 \end{array}
\left[
\begin{array}{cc}
R_{f_1} & \\
K_1\dot\theta & R_a + R_{f_2} - K_2\dot\theta
\end{array}
\right]
\begin{array}{c} 1 \\ 2 \end{array}
\left[\begin{array}{c} i_1 \\ i_2 \end{array}\right]
\end{array}
$$

and that the torque equation is

$$ T = i_2(K_1 i_1 - K_2 i_2). $$

References

[1] BIRD, B. M., McCLOY, K. and CHALMERS, B. J. (1967). New methods of stabilising doubly fed slipring machines. *Proc. IEE*, **114**, 791–796.
[2] KRON, G. (1938). *The application of tensors to the analysis of rotating electrical machinery*. General Electric Review, Schenectady.
[3] HANCOCK, N. N. (1964). *Matrix analysis of electrical machinery*. Pergamon, Oxford.
[4] JONES, C. V. (1967). *The unified theory of electrical machines*. Butterworths, London.
[5] KRON, G. (1959). *Tensors for circuits*, 2nd edition. Dover, New York.
[6] LYNN, J. W. (1963). *Tensors in electrical engineering*. Arnold, London.

Bibliography

The following lists are not exhaustive, and are intended as suggestions only.

Background reading

KIP, A. F. (1969). *Fundamentals of electricity and magnetism,* 2nd edition. McGraw-Hill, New York.
SCOTT, R. E. (1962). *Elements of linear circuits.* Addison—Wesley, Reading, Massachusetts.
STEPHENSON, G. (1961). *Mathematical methods for science students.* Longmans, London.

Further reading

CARTER, G. W. (1967). *The electromagnetic field in its engineering aspects,* 2nd edition. Longmans, London.
DANIELS, A. R. (1968). *The performance of electrical machines.* McGraw-Hill, Maidenhead.
FITZGERALD, A. E., KINGSLEY, C., Jr. and KUSKO, A. (1971). *Electric machinery,* 3rd edition. McGraw-Hill, New York.
HANCOCK, N. N. (1964). *Matrix analysis of electrical machinery.* Pergamon, Oxford.
HINDMARSH, J. (1970). *Electrical machines and their applications,* 2nd edition. Pergamon, Oxford.

Answers to Problems

Chapter 1

1.1 2 980 N.

1.3 The machine will not work.

1.4 Use Ampere's circuital law and the reciprocal property of mutual inductance.

1.5 0.5 H;   7.76 A;   0.193 H;   1 790 N.

1.6 $-\frac{1}{4}L_2 I_m^2 \sin 2\phi$.

1.7 $A = 1.14$; $B = 0.0052$;   hysteresis loss 57 W;   eddy current loss 13 W.

Chapter 2

2.1 90.9%;   1.08 Ω.

2.2 $T = \dfrac{KV^2}{(R + K\omega_r)^2}$.

2.4 (a) 20 A;   (b) 90 rad/s;   (c) 20 A;   (d) 10 A;

(e) 95 rad/s; $\dfrac{d\omega}{dt} + 40\omega = 3\,800$.

Chapter 3

3.2 1:3.

3.4 $v_1 i_1 = v_2 i_2$; $Z_1 = k^2 / Z_2$;   inductance of value $k^2 C$.

3.5 $R_l = 739\,\Omega$; $X_m = 193\,\Omega$; $R_e = 0.127\,\Omega$; $X_e = 0.139\,\Omega$; $n = 3.04$;
(a) 659.9 V;   (b) 97.41%;   (c) 1.38%.

3.6 (a) Excessive magnetizing current will burn out the transformer.
(b) Very low magnetizing current, eddy current loss unchanged, hysteresis loss 0.233 times normal value.

Chapter 4

4.1 $F(0)$ is the displacement of the θ axis to give equal positive and negative areas.

4.5 $T = \frac{1}{2}K^2 I_m^2 \pi r l A_1 \sin\{2(\omega - \omega_r)t + 2\alpha\}$.

Chapter 5

5.1   36 kW;   500 V;   58.3 A;   0.857;   45 kW.

5.3 $f = \dfrac{1}{2\pi}\sqrt{\dfrac{pT_0}{J \tan \delta_0}}$.

5.4   By Lenz's law, the rotor oscillations will be damped.

Chapter 6

6.4   Effective rotor resistance R_2'/s in the first case, $R_2'/(2 - s)$ in the second.

6.5   Two equivalent circuits in series, with element values halved.

APPENDIX A

Air-gap Field Components and the Maxwell Stress

Fig. A.1 shows an idealized machine structure. The stator and rotor surfaces are smooth; the permeability of the iron is assumed to be infinite; and the windings are represented by 'current sheets' of negligible thickness on the stator and rotor surfaces. Current flows in the axial direction, i.e. perpendicular to the

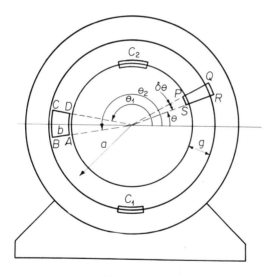

Fig. A.1

plane of the paper. We define a surface current density K to be a vector in the direction of current flow, such that the current in a portion of the sheet is $K\,ds$, where ds is an element of length perpendicular to K. This is analogous to the definition of the volume current density J.

Corresponding to a sinusoidally distributed winding we have a sinusoidal current sheet; the surface current density K varies sinusoidally round the air gap, i.e.

$$K = K_m \cos \theta. \tag{A1}$$

If the density of the stator current is K_1, the corresponding magnetic field may be found by applying Ampere's circuital law to selected paths. The radial component H_{1r} may be evaluated from a path such as $PQRS$ (Fig. A1). The current enclosed is approximately $K_1(\theta) a\, \delta\theta$, where a is the mean radius of the air gap; H_1 is zero along QR and SP because the permeability is infinite; and H_{1r} is assumed constant along a radial path such as PQ. We have

$$K_1(\theta) a\, \delta\theta = \oint H_1 . ds = gH_{1r}(\theta + \delta\theta) - gH_{1r}(\theta), \tag{A2}$$

where g is the radial length of the air gap, which is assumed to be small in comparison with the mean radius a. Thus

$$K_1(\theta) = \frac{g}{a} \frac{\partial H_{1r}(\theta)}{\partial \theta},$$

i.e.

$$H_{1r}(\theta) = \frac{a}{g} \int_{\theta_0}^{\theta} K_1(\theta)\, d\theta + H_{1r}(\theta_0). \tag{A3}$$

Substituting for $K_1(\theta)$ from equation A1 and putting $H_{1r}(\theta_0) = H_{1r}(0) = 0$ gives

$$H_{1r} = \frac{a}{g} K_{1m} \sin \theta, \tag{A4}$$

which is equivalent to equation 4.5 if $aK_{1m} = ik$.

Next we show that there must be a tangential component of magnetic field. Consider a path in the air gap which links no current such as ABCD in Fig. A1. We have

$$0 = \oint H_1 . ds = \int_B^C H_{1s} ds + \int_D^A H_{1s} ds + \frac{b}{g} aK_{1m}(\sin \theta_2 - \sin \theta_1), \tag{A5}$$

where $AB = CD = b$. Equation A5 shows that there must be a tangential component H_{1s} along BC or AD. To evaluate this component, consider the boundary conditions at the stator and rotor surfaces. We take closed contours C_1 and C_2 enclosing lengths δs_1 and δs_2 of the respective surfaces; and we allow the ends of the contours to shrink to zero in such a way that one curved side is just in the iron, while the other curved side is just in the air. For the contour C_1, the current enclosed is $K_1 \delta s_1$, and Ampere's circuital law gives

$$K_1 \delta s_1 = \oint_{C_1} H_1 . ds = H_{1s} \delta s_1, \tag{A6}$$

i.e. $H_{1s} = K_1$ on the air-gap side of the stator surface. Since we are considering the field due to stator current alone, the contour C_2 encloses no current, and we have

$$0 = \oint_{C_2} H_1 \cdot ds = H_{1s} \, \delta s_2. \tag{A7}$$

Thus $H_{1s} = 0$ on the air-gap side of the rotor surface. If H_{1r} is independent of r, and the air-gap length g is small in comparison with the mean radius a, it is readily shown that H_{1s} varies linearly with r from a value of 0 at the rotor surface to K_1 at the stator surface. These results for H_{1r} and H_{1s} agree with the exact solution of the field equations given by White and Woodson [1], subject to the condition that $g \ll a$. A similar argument holds for the rotor field H_2; the component H_{2r} is independent of r, and H_{2s} varies from a value 0 at the stator surface to K_2 at the rotor surface.

Consider the force exerted on an element δs of the rotor when both stator and rotor currents are present. From equation 1.43 the tangential Maxwell stress is

$$t_s = \frac{B_r B_s}{\mu_0} = B_r H_s = B_r(H_{1s} + H_{2s}). \tag{A8}$$

At the rotor surface we have $H_{1s} = 0$ and $H_{2s} = K_2$. If l is the axial length of the element the tangential force is given by

$$\delta F_s = t_s l \, \delta s = B_r K_2 l \, \delta s. \tag{A9}$$

The quantity $K_2 \, \delta s$ is just the current δi in this portion of the rotor surface, so equation A9 becomes

$$\delta F_s = B_r l \, \delta i. \tag{A10}$$

Equation A10 is equivalent to equation 4.18, showing that the Maxwell stress calculation is equivalent to evaluating the force on a current element in a magnetic field.

Reference

[1]   WHITE, D. C. and WOODSON, H. H. (1959). *Electromechanical energy conversion*. Wiley, New York.

APPENDIX B

Three-phase to Two-phase Transformation

We consider the conditions under which it is possible to replace a three-phase machine winding with a two-phase winding, so that the performance of the machine is unchanged. The two windings are shown diagrammatically in Fig. B1, and they are assumed to be sinusoidally distributed. Let currents i_a, i_b,

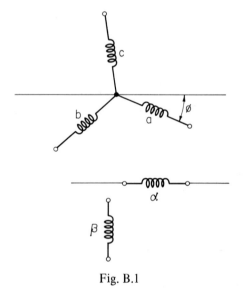

Fig. B.1

i_c flow in the three-phase winding; then the m.m.f. at an angle θ from the reference axis will be

$$F_3 = k_3\{i_a \cos(\theta + \phi) + i_b \cos(\theta + \phi + 2\pi/3) + i_c \cos(\theta + \phi + 4\pi/3)\}. \quad \text{(B1)}$$

If currents i_α, i_β flow in the two-phase winding, the corresponding m.m.f. will be

$$F_2 = k_2\{i_\alpha \cos\theta + i_\beta \cos(\theta + \pi/2)\}$$

$$= k_2(i_\alpha \cos\theta - i_\beta \sin\theta). \tag{B2}$$

The two windings will be equivalent if $F_3 = F_2$ for all values of θ. Equation B1 may be expanded to give

$$F = k_3 i_a(\cos\theta \cos\phi - \sin\theta \sin\phi) +$$

$$+ k_3 i_b\{\cos\theta \cos(\phi + 2\pi/3) - \sin\theta \sin(\phi + 2\pi/3)\} +$$

$$+ k_3 i_c\{\cos\theta \cos(\phi + 4\pi/3) - \sin\theta \sin(\phi + 4\pi/3)\}, \tag{B3}$$

and the condition for equivalence is obtained by equating the coefficients of $\sin\theta$ and $\cos\theta$ in equations B2 and B3. This gives

$$\left.\begin{aligned} i_\alpha &= \frac{k_3}{k_2}\left\{i_a \cos\phi + i_b \cos(\phi + 2\pi/3) + i_c \cos(\phi + 4\pi/3)\right\}, \\[2mm] i_\beta &= \frac{k_3}{k_2}\left\{i_a \sin\phi + i_b \sin(\phi + 2\pi/3) + i_c \sin(\phi + 4\pi/3)\right\}. \end{aligned}\right\} \tag{B4}$$

These relations may be written in matrix form:

$$\begin{bmatrix} i_\alpha \\ i_\beta \end{bmatrix} = k \begin{bmatrix} \cos\phi & \cos(\phi + 2\pi/3) & \cos(\phi + 4\pi/3) \\ \sin\phi & \sin(\phi + 2\pi/3) & \sin(\phi + 4\pi/3) \end{bmatrix} \begin{bmatrix} i_a \\ i_b \\ i_c \end{bmatrix} \tag{B5}$$

where $k = k_3/k_2$. The corresponding relationship for voltage may be deduced if we stipulate that the total instantaneous input power is to be the same for both windings. We write equation B5 in the form

$$i_2 = A\, i_3 \tag{B6}$$

where i_2 is the column matrix of two-phase currents, and i_3 is the matrix of three-phase currents. By an argument similar to the one used in section 7.5, the corresponding voltages must be related by the matrix equation

$$v_3 = A^T v_2. \tag{B7}$$

Since the matrices A and A^T are singular, equations B6 and B7 cannot be inverted as they stand. For a machine with no neutral connection, however, the three currents i_a, i_b, i_c are not all independent; we must have

$$i_a + i_b + i_c = 0, \tag{B8}$$

showing that there are only two independent three-phase currents. For machines with balanced windings, it may be shown (Jones [1]) that when there is no neutral connection the three phase voltages v_a, v_b, v_c are also related by the equation

$$v_a + v_b + v_c = 0. \tag{B9}$$

Equation B9 is exactly true for machines with a uniform air gap, and a reasonable approximation for salient-pole machines. We may incorporate equations B8 and B9 in equations B6 and B7 by including additional elements in the matrices:

$$\begin{bmatrix} i_\alpha \\ i_\beta \\ 0 \end{bmatrix} = k \begin{bmatrix} \cos\phi & \cos(\phi + 2\pi/3) & \cos(\phi + 4\pi/3) \\ \sin\phi & \sin(\phi + 2\pi/3) & \sin(\phi + 4\pi/3) \\ m & m & m \end{bmatrix} \begin{bmatrix} i_a \\ i_b \\ i_c \end{bmatrix}, \tag{B10}$$

$$\begin{bmatrix} v_a \\ v_b \\ v_c \end{bmatrix} = k \begin{bmatrix} \cos\phi & \sin\phi & m \\ \cos(\phi + 2\pi/3) & \sin(\phi + 2\pi/3) & m \\ \cos(\phi + 4\pi/3) & \sin(\phi + 4\pi/3) & m \end{bmatrix} \begin{bmatrix} v_\alpha \\ v_\beta \\ 0 \end{bmatrix} \tag{B11}$$

In these equations m is an arbitrary constant, and the factor $k = k_3/k_2$ can be assigned any value, since k_3 and k_2 are arbitrary. Since the square transformation matrix is non-singular it can be inverted; if we write equations B10 and B11 in the form

$$i_2' = A' i_3, \tag{B12}$$

$$v_3 = A'^T v_2', \tag{B13}$$

then the required inverse relationships are

$$i_3 = (A')^{-1} i_2', \tag{B14}$$

$$v_2' = (A'^T)^{-1} v_3. \tag{B15}$$

It would be very convenient if voltage and current transformed in the same way, i.e. if

$$A'^T = (A')^{-1} \tag{B16}$$

and $$A' = (A'^T)^{-1}. \tag{B17}$$

We thus require the matrix A' to be orthogonal, and this will be the case if $k = \sqrt{(2/3)}$ and $m = 1/\sqrt{2}$. Finally, we may omit the additional term from the matrices and for convenience let $\phi = 0$. Then

$$\begin{bmatrix} i_\alpha \\ i_\beta \end{bmatrix} = \sqrt{(2/3)} \begin{bmatrix} 1 & -1/2 & -1/2 \\ 0 & \sqrt{3}/2 & -\sqrt{3}/2 \end{bmatrix} \begin{bmatrix} i_a \\ i_b \\ i_c \end{bmatrix}, \tag{B18}$$

$$\begin{bmatrix} i_a \\ i_b \\ i_c \end{bmatrix} = \sqrt{(2/3)} \begin{bmatrix} 1 & 0 \\ -1/2 & \sqrt{3}/2 \\ -1/2 & -\sqrt{3}/2 \end{bmatrix} \begin{bmatrix} i_\alpha \\ i_\beta \end{bmatrix}, \qquad (B19)$$

with similar equations for the phase voltages.

We can thus convert from three to two phases and vice versa for all conditions of operation — balanced or unbalanced, steady-state or transient — provided only that there is a three-wire connection to the three-phase winding. A three-phase machine may therefore be analysed in terms of the simpler two-phase model, and the results transformed back to three phases by means of equation B19. A particular advantage of choosing $\phi = 0$ is that i_a is independent of i_β, and is simply equal to $\sqrt{(2/3)}i_\alpha$.

Reference

[1] JONES, C. V. (1967). *The unified theory of electrical machines.* Butterworths, London.

Index